# DEVELOPING EFFECTIVE SAFETY SYSTEMS

*Ian G Wallace*

*Safety consultant, formerly employed by*
*Du Pont, Phillips Petroleum, Conoco and Neste Oy*

INSTITUTION OF CHEMICAL ENGINEERS

**Published by**
**Institution of Chemical Engineers,**
**Davis Building,**
**165–189 Railway Terrace,**
**Rugby, Warwickshire CV21 3HQ, UK.**
IChemE is a Registered Charity

© 1995 Ian G Wallace

ISBN 0 85295 358 5

Printed in the United Kingdom by Galliard (Printers) Ltd, Great Yarmouth

*To Victoria, Nicola, Ceri, Katie and April*

# PROLOGUE

This book is based on my experiences collected over more than 30 years in industry. I have worked in a wide variety of industries and sizes of companies, from nuclear energy to oil and gas exploration and production, from large multi-national companies (now referred to as global companies) to a small self-owned partnership. Initially I was involved in production supervision, but moved into the safety field full time in 1978.

I have based this book on my gradual awakening to what I believe are the key factors in achieving a high standard of safe operations. The experiences which really crystallized my ideas were gained working first for Du Pont de Nemours in their UK company, Phillips Petroleum Ltd, then for Conoco UK and finally for Sovereign Oil and Gas Plc before and after its takeover by Neste Oy (the Finnish State oil and chemical company). At Sovereign Oil and Gas I had the opportunity to set up safety policies, procedures and systems effectively from scratch. Since then I became a consultant with EWI, a major international safety and environmental consultancy operating in the nuclear, energy and oil and gas industries, and recently a self-employed consultant, and have had a chance to see many forms of safety management and compare their effectiveness.

I must emphasize that while this book is based on my own experiences, the contents consists of my own opinions and ideas and not the policies and procedures of any of the companies that have employed me.

I have written the book assuming that the reader is starting from scratch in setting up a safety programme, so there is a progressive build-up of policies, systems and procedures. Of course this is an ideal, and most unlikely, situation. Almost inevitably, you will be joining an organization, or already be part of an organization, with an existing set of policies, procedures and programmes, and will wish to develop the systems further to achieve a higher level of performance. Even if you are joining a new organization, the pressures to get moving will normally require that a number of procedures be put in place immediately. If there are existing policies and procedures, they can continue until replaced by the relevant section of the new regime. The need to introduce new procedures immediately is more of a problem as it is essential to have certain systems up and running in the short term. In this situation you will have to identify the critical procedures and put them in place immediately. And with any luck you will be able to discern the shape of the policies the organization requires, because the procedures you introduce will have to be in line with ultimate objectives.

Ian G Wallace

**ACKNOWLEDGEMENTS**
My sincere thanks must go to my wife who put up with me through the traumas involved in writing and correcting this book, to Dr Barry Thorne for his help and advice and to Julie, Jenny, Lisa and Tracy who valiantly coped with my scribbles and attempts at word processing.

**NOTE TO READERS**
There are a number of references in this book to UK legislation and organizations. No attempt has been made to provide similar information for other countries; readers working outside the UK will undoubtedly be acquainted with the relevant equivalents and make suitable allowances.

# CONTENTS

# 1.    COMMITMENT AND CULTURE

Many people ask companies which are successful in achieving high standards of safety, 'what special technique do you use to achieve such good performance?'. This myth should be exploded once and for all — 'there is no magic cure for lack of safety'. Successful companies *do not* have a secret safety technique or programme. They generally operate very similar equipment to the companies with poor records. After all, a circular saw has just as many teeth on it whether it is operated by company A or company Z. The real secret is sheer hard work and total commitment by everyone. As the tourist visiting one of the more ancient Oxford colleges was told when he asked how the staff achieved the superbly manicured lawn — 'Oh there's no magic, you just have to cut it and roll it regularly for about 600 years'.

In a way that story can be related to achieving a high standard of safety. Although it doesn't take 600 years to accomplish something significant, there is something very similar about the patience and constant dedication that is required. The commitment must be made by everyone, from the Board of Directors to the part-time office cleaners.

To start this process there are three fundamental decisions that a company has to take:

(1)  The company has to dedicate itself (and 'dedicate' is a key word here) to establishing a policy and culture that looks at safety in the same way as it looks at production, profits, research, planning, sales and all the other key ingredients of a successful business.

(2)  The company management has to build the framework that gets the safety policy flowing effectively through the entire company.

(3)  The safety issue has to be slotted into the career structures, goals and objectives, rewards systems and all the other processes that are essential to keeping the business moving along. In other words, the company must value and recognize safety achievement in a way that lets everyone know that it is serious about safety.

Having done all that, one other thing is crucial: the company priorities must be recognized and safety put right at the top. This may turn out to be the best management decision the company ever takes. It is good for morale, good for people, good for productivity, good for profits and basically good for everyone with a stake in the company.

There are three very sound reasons why safety should be such a central concern. First, and most important, we as human beings cannot accept that people are killed and maimed in our operations, especially as almost all accidents can be avoided. Second, society demands that every effort is made to operate safely and, if this is not achieved, the regulatory authority will take steps to ensure that standards are raised. And third, it is good business to be safe.

However cynical, hard-nosed, results-orientated, bottom-line-dedicated the company may be ... a good argument has yet to emerge that challenges the basic principle that good safety practice is plain good business.

It is clear that there is a direct link between good safety management and good business management. Companies with high safety standards are those with high operational standards. Indeed, this is not surprising, because the management actions necessary to achieve safety are exactly those required to achieve business efficiency.

The tools that managers use to improve safety are the same ones that they use to improve any aspect of business — setting goals, monitoring progress, questioning decisions, probing dark corners and listening to staff at all levels. For example, operating procedures are one of the keys to safe operations and managers must ensure that critical procedures are identified and written down. But developing good procedures and putting signatures to glossy documents is just the beginning. Next, managers must ensure that the procedures are implemented.

One of the tools in achieving safety is a formal comprehensive programme of safety audits — ranging from periodic in-depth management and technical audits to frequent supervisory inspections. Management must regularly visit operational sites — observing, checking and questioning. Perhaps the operators have never received the procedures or been trained in their use. Perhaps the procedures were written behind a desk by someone unfamiliar with the practicalities of the operations. Perhaps the operators know and understand the procedures, but have formed the impression that it does not really matter if they are observed. Only committed managers will find and rectify these problems.

One initiative that can be adopted is 'unsafe auditing', as a means of reducing unsafe acts and unsafe situations. The technique, pioneered by the chemical company Du Pont, teaches supervisors and staff to spot safety problems and communicate them to other workers — with the aim of creating a culture where all staff constantly monitor their own and their colleagues' safety. If the technique is to succeed in raising staff safety commitment, it must rigorously avoid any 'policing' connotation — reporting safety violations for disciplinary purposes — and concentrate on positive counselling.

Three key areas of safety management are communication, training and monitoring.

## COMMUNICATION

Communication starts with senior managers, who must be willing to devote sufficient of their time and attention to the subject that they convince the organization of the importance of safety. They must demonstrate that their commitment to safety is just as great as their commitment to clinching that deal, reaching production targets and so on.

However, management exhortation will not in itself bring about improvement. Managers have to bring about a change in the way the organization works so that supervisors at every level are communicating safety directly to their subordinates.

This communication is not just one-way, spreading the 'safety gospel' from on high. Rather, two-way channels must be established so that the safety concerns and suggestions of workers are communicated back up the line — where they must be, and be seen to be, acted on. Nor must this communication be separate from the ordinary work communication, attended to only at periodic safety meetings and then forgotten. It has to be a continuous integral part of the daily work routine.

## TRAINING

Of course, if managers and supervisors are going to perform this vital role properly they need training. They must be trained in safety management and learn how to identify unsafe acts and situations and to correct them. Managers also need training in analysing accidents so that they can look behind the obvious causes and identify areas of basic safety weakness.

## MONITORING AND AUDITING

As in all areas of management, performance monitoring is essential for improvement. Management must insist that safety performance is one of the first items discussed in the report of any operating unit — whether the annual report of a major operating company or the weekly report of a section. At a personal level, safety performance must be an integral part of the performance assessment carried out for all staff appraisals.

Audits are a vital way of verifying that a company's safety management is working properly. However, to be successful they must be viewed as part of the 'learning' process rather than as a 'police' action. Technical audits check both the technical integrity — the hardware — and the safety procedures — the software — of a facility, since it is only by marrying these two aspects that safety can be achieved.

Accident causation models suggest that at least 98% of accidents are caused by human failing, management, supervisors, employees or designers. Thus, to have a high standard of safety performance, people must do things correctly. This cannot be achieved by coercion, threats or use of discipline. It can only come about when everyone has a high level of safety awareness; in fact it is necessary, for really high standards, to achieve an instinctive, rather than deliberate, awareness of safety. In other words it needs a 'safety culture'.

Having achieved commitment and generated the correct 'safety culture', the means to achieve safe operation are also required. This can only come about when the causes of accidents and hazards are understood.

Psychological researchers who have studied accident causation and safety improvement programmes see a distinction between 'active failures' by operational staff (production operators, drillers, pilots, drivers and so on) and 'latent failures', which are in the system as a result of the decisions of planners, designers and managers. They argue that while active failures trigger accidents, they normally do so only because of latent failures already embedded in the system.

An example of the distinction between the two types of failures may be seen in newspaper reports some years ago of an aircraft which crashed on a motorway in England. These reports suggested that a fault occurred in one engine, but that the pilots switched off the other engine by mistake. That would clearly be an active failure, but if two experienced pilots made such a mistake, it seems likely that a number of latent failures were already present in areas such as instrument design, cockpit procedures and training.

The researchers also looked into the psychological reasons for failures and analysed a number of major disasters. They found a common thread of:
- poor safety motivation;
- inadequate defences against mistakes;
- poor operating and maintenance procedures;
- inadequate training;
- bad organization;
- lack of communication;
- priority of commercial goals over safety.

All of the disasters examined could, quite simply, be attributed to failures of management. If this analysis is correct, safety improvements must be sought throughout an organization by removing both latent and active failures.

Most safety professionals agree that there are three basic requirements for safety:

- good quality design and engineering;
- well thought-out and pragmatic procedures;
- people with the motivation and understanding to work safely.

Safe design principles are documented in extensive guidelines and standards — often with legal force — and to a significant extent are uniform throughout specific industries. There is also considerable agreement within industry on key procedures such as work permit systems.

Above all, it is vital that everybody understands the priority that managers give to safety and their determination not to allow operations without appropriate safety systems in place. Difficult tasks, no doubt, but it is what managers are paid to do.

It is essential that managers and supervisors are held personally responsible for the safety of operators under their control — whether company staff or contractor. It is not acceptable for a manager or supervisor to plead that the cause of an accident was the failure of a subordinate.

The main cause of unsafe operations was best expressed by a British judge investigating a major disaster. 'From top to bottom', he said, 'the body corporate was infected with the disease of sloppiness'.

Everyone knows what sloppiness is — things poorly designed, poor quality control, procedures habitually short-circuited, poor communications and bureaucratic delay. A sloppy company is not only an unsafe company, but also an inefficient company.

Morally, there is a duty to be safe. It is unacceptable for people to be killed or maimed in operations while accidents can be prevented. It is also good business to be safe, and companies that are well managed are likely to be both safe and prosperous. Everyone can work for that.

The principles of effective safety management are as follows:

- management demonstrates a visible commitment to safety;
- safety is one of the responsibilities of line management;
- competent safety advisers are appointed;
- high safety standards are available and well understood;
- audits of safety standards and practices are carried out;
- effective safety training is conducted;
- sound safety policies are in place;
- realistic safety targets and objectives are set;
- techniques to measure safety performance are applied;
- injuries and incidents are investigated thoroughly and followed up;
- effective motivation and communication is inherent in all activities.

5

But it is essential to realize that it is possible to 'safety a company out of business' just as easily as it is to be put out of business by a major incident. It is necessary to ensure that the safety objectives are practical and assist in achieving an efficient operation. If a bureaucratic, over-complicated system is instituted it snarls up the operations and prevents anyone doing anything productive. Safety excellence cannot be achieved overnight. It takes time to generate the commitment, even longer to develop the culture and a lifetime to introduce the programmes needed to ensure safe operations — bearing in mind that it is necessary to introduce new or modified programmes regularly in order to maintain interest and awareness, and hence commitment, and thus the continuance of 'the culture'.

# 2.  PRINCIPLES, STANDARDS, POLICY AND PERSONNEL

**SAFETY, HEALTH AND ENVIRONMENTAL PROTECTION PLAN**

The key to success in any business is the policy and decision-making process. Successful organizations are usually found amongst those with clear and unambiguous objectives supported by an effective management which directs and controls the organization. A successful safety programme requires the application of conventional management systems and techniques. It must be seen as part of the normal management of the company, making demands on management time and competing for available resources and cash in the same way as other relevant factors such as research and development, new machinery, wages and salaries and quality control. To achieve this it is necessary to prepare a plan. Figure 2.1 overleaf illustrates the structure of the safety, health and environmental protection plan that was developed for a small UK company. The development of a plan suited to a particular company allied to the commitment of senior management assists in the generation of a 'safety culture'.

It is not possible to create the right culture overnight, but if it is not worked on it will never be achieved. First, the Managing Director decides what the safety objectives are. It is difficult to justify aiming to be 'the best' but it is certainly practicable to be 'above average' or, better still, 'in the upper quartile'. Once the goals are set and the Managing Director really committed, the Safety Manager can help to achieve those goals by turning them into a set of safety and environmental principles or safety ethics backed up by relevant standards. Everyone in the company must be motivated to strive to attain the standards, and their performance should be measured and compared against them.

Once the principles and standards have been developed and agreed by the Managing Director they must be approved by the Board of Directors as official company policy, signed by the Managing Director and issued by the Executive Officers to all company units. As a corollary to the adoption by the Board, the company Safety Manager is expected to give regular reports to the Board identifying how the company is progressing towards the attainment of the standards.

Safety, health and environmental principles and standards are, of course, only a start. They need turning into a series of policies, supported by appropriate programmes and procedures as shown in Figure 2.1:

- a safety and health programme;
- a work force involvement programme;
- a training programme;
- an accident and incident programme;
- an emergency response programme;
- an environmental protection programme.

Each of these programmes is multifaceted; the topics relevant to safety are discussed in detail in the following chapters.

The safety, health and environmental protection plan must be developed further if the company operates internationally or in a variety of disparate industries. In this case what is suitable for one location may not be acceptable in another and yet all the policies and procedures, no matter how diverse, must hang together and be in line with corporate objectives and policies to form a cohesive whole.

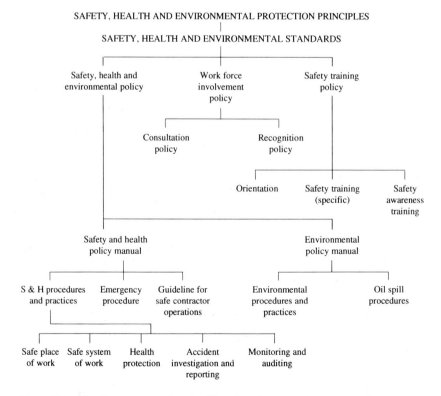

Figure 2.1    Small company safety, health and environmental protection plan.

CORPORATE SAFETY, HEALTH AND ENVIRONMENTAL PROTECTION PRINCIPLES

CORPORATE SAFETY, HEALTH AND ENVIRONMENTAL POLICY

Division A          Division B          Division C

Safety, health and environmental protection principles

Guidelines on safety, health and environmental protection policy

Facility A          Facility B          Facility C

Safety, health and environmental protection management manual

S & H standards          Environmental protection standards

S & H programme          Environmental programme

Figure 2.2    Corporate safety, health and environmental protection plan.

The company made up of different divisions requires a set of safety, health and environmental protection principles which are elaborated into a policy as shown in Figure 2.2. These will be very general and capable of application throughout the company. But of course they will be too general to be of much use to managers operating in very different fields. Thus each division needs to take the principles and develop a specific divisional set which apply the corporate principles to its field of activity. The corporate policy along with the divisional principles will allow divisional guidelines on the corporate policy to be produced. Note that these guidelines are not policy, since the overall corporation should have only one safety, health and environmental protection policy. They merely interpret the company policy and explain how it applies to the division's activities. The division may well operate in a variety of different parts of the world with very different standards, infrastructures and current capabilities. Or it may carry out very different types of operation in different locations with a variety of hazards. Thus each location or facility as appropriate needs to develop a safety, health and environmental protection management manual and standards which lay down how their particular business will be managed with respect to safety, health and the environment, the programmes that will be used and the standards that will be applied.

9

Figure 2.3 shows how the facility management manual and standards can be developed into the necessary programmes to achieve the desired performance.

## SAFETY, HEALTH AND ENVIRONMENTAL PRINCIPLES

The company principles set the tone of the company and define the way that it carries out its business. They need to address:

(1) The level of importance attached by the company to safety, health and environmental protection.

(2) The overall level of safety, health and environmental performance that the company wishes to achieve.

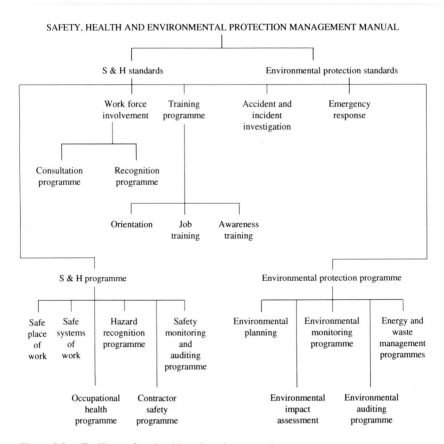

Figure 2.3    Facility, safety, health and environmental programme.

(3)  The standards of safety, health and environmental protection that are to be used to judge company performance.

(4)  The quality assurance standards that are to be applied to safety, health and environmental activities.

(5)  The general responsibilities for safety, health and environmental protection of the Board, management, supervision and employees.

(6)  The company attitude to safety, health and environmental training.

(7)  The level of importance attached by the company to employee involvement and participation.

## SAFETY, HEALTH AND ENVIRONMENTAL STANDARDS

The safety, health and environmental standards define the company standards to be applied to the following:

(1)  The design and construction of all new facilities.

(2)  The design of modifications to existing facilities.

(3)  The action taken to eliminate hazards which could lead to incidents.

(4)  The action taken to eliminate hazards which could lead to an injury, industrial illness or release of a pollutant to the environment.

(5)  How the company will ensure that safe systems of work are available for all work being carried out under its control.

(6)  How the company will ensure that all personnel — company and contract — are adequately qualified and experienced to allow them to carry out their work safely and without risk to the environment.

(7)  The safety, health and environmental training which will apply to all personnel, company and contractor.

(8)  What hazard studies will be carried out on all new plant and process designs to identify hazards and assess their consequences.

(9)  The procedures to ensure that no modification to plant, processes, facilities and software compromises safety, health or environmental standards.

(10) The arrangements to ensure that emissions to the environment are monitored, are kept as low as reasonably practicable and do not exceed legislative limits.

(11) The arrangements to ensure that the exposure of personnel to chemical, physical and biological agents is monitored and controlled below the legislative limits.

(12) The priority to be applied to substitution or engineering controls over controlled access to prevent exposure of personnel to substances, and the place of personal protection.

(13) Any medical standards to be applied to personnel and any substance abuse policy to be applied to company operations.

(14) The incident, near miss and hazard reporting and investigation procedure.
(15) The arrangements to be put in place to deal with emergencies and ensure that personnel are ready to cope with any that arise.
(16) The audit programme to be applied to all company activities and operations.


## POLICY

The UK Health and Safety Executive (HSE) and others have published a number of reports and booklets on the subject of company safety and health policy. They all agree that a policy should cover at least the following:

- organization and responsibilities of all levels in the company;
- involvement and consultation;
- training;
- how a safe place of work is ensured;
- arrangements to achieve a safe system of work;
- the programme to ensure health protection;
- accident and incident reporting and investigation;
- the emergency response procedures;
- the inspection and audit programme;
- people.

The essential need is for all the elements involved in health, safety and environmental protection to receive correct analysis, consideration and appraisal within the routine of business management, and for it to be seen as of equal importance and an integral part of the corporate system. The Chief Executive and the Board of Directors are the persons who give direction and purpose to any organization. They are therefore responsible for the 'business decisions'; in order to make such decisions they have to receive and interpret a variety of relevant, interacting and irrelevant information. They are responsible for leadership and direction and ensuring that there is an adequate supporting infrastructure which sustains the system and provides effective communication.


### THE CHIEF EXECUTIVE

The key person is the Chief Executive, who must understand and recognize the importance of safety, health and environmental protection so that proper judgements may be made and clear policies generated. The Chief Executive has the prime responsibility and is the ultimate custodian of authority. This means ensuring that there is an effective monitoring system in place and working, to allow confirmation that agreed actions are being implemented and progress is being made towards the company objectives and goals. The monitoring system starts with the company Executives; Executives are all held accountable for their actions or lack of action. The activities and attitudes of the Executive inevitably determine and shape the attitudes and responses of line management and hence the real shape and content of company policy, as opposed to what the paperwork says. If the Chief Executive allows the programme to degenerate or not to be enforced, both the Chief Executive and the programme lose credibility and the respect of the work force. This means that safety and environmental protection are ignored and no amount of discipline or cajoling by supervision will persuade employees to conform to the rules. The ultimate result of this is injuries or even one or more fatalities. The Chief Executive must succeed — there is no acceptable alternative!

## THE SAFETY MANAGER

Since management is dependent on the quality of the information it receives, it is essential that the company Safety Manager provides an adequate advisory service leaving executive decisions to line managers. The Safety Manager's success depends upon an ability as a communicator and being able to get on with employees at all levels. In providing sound information and advice, the Safety Manager helps line management to make correct decisions and take correct actions in matters relating to safety, health and environmental protection. Thus the Safety Manager must be involved in all project work from the outset and follow through to installation, commissioning and operation, including the training of operators.

It is also necessary for the safety professional to be properly qualified and be given sufficient status in the organization, in addition to possessing the appropriate personality to generate charisma, to enable the organization to achieve its desired goals.

The Safety Manager's role includes (but is not limited to) the following:
- monitoring the implementation of the organization's safety, health and environmental protection policy, reporting regularly to the Chief Executive on progress;
- advising line management to help them adhere to the safety, health and environmental protection policy and meet their responsibilities in such matters as:
— design and correct use of plant, equipment and tools;
— identifying unsafe plant working conditions and practices;
— carrying out inspections and making recommendations for correcting defects found;
- assisting in the formulation and implementation of safe systems of work and emergency procedures;
- recommending suitable protective devices and equipment;

- checking compliance with statutory requirements, codes of practice and guidance relevant to the operation;
- ensuring that the necessary registers, records and accident/incident books are being kept properly;
- informing line management of all new legislation, regulations, codes of practice and guidance relevant to the organization and assisting in generating a company response and action plan;
- being a member of the safety, health and environment committee;
- recommending appropriate training programmes to develop safety and environmental awareness at all levels and specialist knowledge as appropriate;
- disseminating information on accidents and incidents which occur within and without the company to increase the knowledge of accident prevention techniques;
- assessing the hazards of materials used and planned to be used, recommending the correct protection and procedures to be used and monitoring the exposure of personnel;
- monitoring adherence of contractors to the company safety policy and procedures;
- assisting in the investigation of accidents, incidents and near misses to establish the cause and recommend actions to prevent a recurrence;
- providing meaningful accident and performance-related statistics;
- ensuring that there are effective emergency response procedures available, personnel are adequately trained and that the company is at an acceptable level of emergency readiness;
- liaising with other departments on safety, health and environmental protection;
- liaising with external bodies such as the regulatory authority and industry organizations;
- keeping abreast of developments and techniques in the fields of safety, health and environmental protection.

In fulfilling these roles the Safety Manager works very closely at all times with senior and line management and employees with the object of ensuring a safe, healthy and environmentally acceptable work place. Of course it is not always possible for one person to do all these things alone and, depending on the size of the organization, the Safety Manager may well have some subordinates to assist. In almost every situation, however, there is one invaluable source of help. As the Robens Committee (which investigated the state of occupational safety and health in the UK and led to the development of the UK Health and Safety at Work etc Act 1974) said, real progress in the promotion of safety will be almost impossible without the full co-operation and commitment of all employees and the participation of their representatives. Therefore an important source of help is the company Safety Representatives and the Safety Committee. By involving the people at the sharp end who have knowledge of what really goes on and a personal interest in what happens, it is possible to identify the real problems more easily and develop realistic solutions. By encouraging their participation and involvement, the commitment of the work force to the policies and procedures is more likely to be obtained. This leads to the real implementation of policies and procedures in the work place, thus developing the 'safety culture' and encouraging the desired level of performance.

# 3.   SAFETY TRAINING

Having obtained a clear commitment that the prime responsibility for safety rests with the Chief Executive, agreed a safety ethic, assembled a set of safety principles and standards and issued a policy, the foundations are in place. But before people can actually achieve anything they need to know how to do it. This can only be done by training. The company therefore needs a safety training policy.

It might well be said, 'but that is the province of the Training Department, not the Safety Department'. In a sense this is true. But the Training Department cannot be expected to be knowledgeable in all the detailed expertise of every single department in the company. The Training Department needs help, and since safety spans the whole of the company and is involved to some extent in all aspects of training, the Safety Department should control the safety training policy and standards, with the Training Department looking after the implementation and administration of the programme.

## SAFETY TRAINING POLICY

Company policy should indicate that training is an integral and important part of the overall safety and health policy and that safety training is a normal part of all vocational training. The responsibility for ensuring that personnel are properly trained and competent to do their jobs safely rests with line management. Thus each manager determines who needs to be trained in what. However, the company should lay down minimum standards of training aimed at ensuring that personnel are aware of the hazards associated with their work, the control measures necessary to minimize the risk of personal harm, or loss or damage to property or the environment. The policy should require contractors to be trained to equivalent standards.

The training standards define the minimum safety training required for each type of position in the company. The document also specifies suitable training courses which meet the standard. Obviously the list will not be exclusive and provision should be made to allow other training and/or courses to be considered and accepted or rejected as appropriate.

The training standards are designed to inform all concerned who should be trained and the standard they should reach, thus ensuring uniform and consistent implementation. Managers also have a responsibility for carrying out periodic audits to ensure that the safety training policy is being adhered to by all personnel, including contractors.

## SAFETY TRAINING OBJECTIVES

The first objective of safety training in industry is an appreciation of personal responsibility for safety, by everyone from senior management to the newest employee. Responsibility for the safety of employees and safety training rests with management at all levels. A further objective is to develop within the company a climate in which everyone is safety conscious and acts safely.

## LEGAL REQUIREMENTS

All employers have a general common law duty to provide a safe place of work, to provide proper plant and equipment, to organize and maintain a safe system of work and to employ competent people. In the UK, statutory requirements for the protection of persons employed are imposed by the Health and Safety at Work etc Act 1974 and subsidiary legislation. The duty to train and supervise employees efficiently is implicit in both criminal law and civil law, including the provision of training, instruction and information to employees.

## MANAGERS AND SUPERVISORS

The basic objectives of management safety training are to make managers aware of the benefits of a high standard of safety and how to achieve that standard.

Train supervisors to understand that the safe operation and maintenance of plant and the efficient management of people requires that correct work procedures are established, understood and adhered to. Ensure that people are allocated to jobs for which they are mentally and physically suitable.

Safety training for supervisors includes knowledge and appreciation of safety policy and standards, hazard recognition, accident prevention and investigation, methods of communicating with and motivating their work force and relevant statutory regulations.

## SAFETY MANAGER

A key part of any effective safety programme is the selection as Safety Manager of a person of the right calibre who is able to advise management. The training of such a person must be related to the kind of job to be done, the number of persons employed and the nature of the processes and materials handled. In smaller companies a senior manager may be appointed Safety Officer and performs this function as part of normal work. But the manager still needs to be properly trained.

## EMPLOYEES

Safety training with clear objectives should be part of all job training. The majority of accidents take place on the shop floor. Thus, much safety training, including fire prevention training, takes place on the job and is of short duration. Fire fighting and first aid training is again of short duration but is normally done off the job. Job/task analysis identifies specific safety training needs. The safety aspects of the analyses should be kept under continuous review, and retraining undertaken when the nature of a task changes.

Operating instructions should cover the safe use of materials, plant and equipment, and the maintenance and use of tools and instruments. Before commencing work, give employees instruction on the hazards of the equipment they will be using, the properties of the materials they have to handle, the risks arising from any gases or fumes, and on how to endure safe operations. In particular, where permits-to-work are to be used for working on plant or equipment, entry to vessels or in other hazardous circumstances, safety training must be carried out to ensure that the procedures are practised and understood. Employees also need to be trained in the action required in an emergency, including the use of any special protective equipment.

## INDUCTION TRAINING

The acceptance of safe working principles and practices by new entrants into a company, particularly those coming to their first jobs, is of paramount importance. An element of safety training should be included in all training programmes for young persons.

The planning stage of any induction or change of job training programme is critical. It is then essential that the devised plan is followed through. Brief all members of staff involved thoroughly about the role they are to play in the programme.

Design induction training to:

(1) familiarize new employees with the industry, the company and the department in which they are to be employed; and

(2) enable them to work safely, effectively and efficiently.

Part of the importance of induction training lies in the fact that it gives new employees a basic understanding of their environment, and helps to provide the confidence and the motivation which are essential prerequisites of further successful training.

Research has indicated that induction training may assist in reducing the rate of labour turnover and increase an employee's productivity and job satisfaction. Also the inclusion of health and safety as part of induction training undoubtedly minimizes the accident rate.

Induction training forms an important part of any formal training programme and should be given to all employees. Do not exclude part-time employees from the programme. Induction training is especially important in the case of young people entering industry for the first time and for employees returning to work after a long break. When people change jobs, check to see if they need further induction into other parts of the organization in which they may now become involved.

When new entrants join a company it is important that they be made to feel at home as quickly as possible. To do this, introduce them systematically to their new surroundings. Each company should adapt the subject matter to suit its own particular needs, but it is essential that, irrespective of size, the programme of induction training is well-planned and presented in an interesting, logical and meaningful manner.

In larger companies, induction normally takes the form of special courses which new employees undergo on arrival or soon afterwards. Small companies, with an insufficient intake of new employees to justify the establishment of formal induction courses, might find it necessary to organize induction on an *ad hoc* and possibly less formal basis, as the occasion demands. The way in which induction training is organized depends on the circumstances of the company and the resources it has available.

If a company has a Personnel Officer or Training Officer, this person should have the responsibility for planning and carrying out induction. However, senior managers, and especially the new entrant's supervisor, should be involved as much as possible in the programme. The role of the employee's supervisor is a vital one and determines, to a very great degree, the efficiency of the training. The supervisor must be convinced of the importance of taking part in induction; this can be greatly aided if the supervisor is involved in the process of developing the programme. Specialists, such as the Health and Safety Advisor or Fire Officer, should be involved where appropriate in the induction programme. Within this framework it is important to allocate time and responsibility for specific coverage of each item of the induction programme for each trainee.

The newcomer to any job is usually unaware of the dangers which can arise in the course of work. It is, therefore, essential that everyone is given basic safety information, including information about:

- national safety legislation;
- the company's safety, health and environmental principles, standards and policy and how they affect the employee;
- safety and emergency procedures;
- major hazards and risks.

If induction is carried out formally these points can be included in the course; otherwise include them in the supervisor's check list of information to be given to new starters. Table 3.1 gives a sample check list.

Young employees should be given detailed explanations of why certain practices are safe and others unsafe. The extra time and effort spent on the safety training of young people is very worthwhile, for this is the most formative period of their working lives. If a reasonable approach to their own safety and that of others is developed early, it is likely to remain throughout their careers.

Whenever possible make full use of learning aids (programmed texts, slides, video tapes, films, recorded commentaries, publications and so on) to assist the process of induction. Handbooks should be issued if available, but they are not a substitute for induction training.

It is important that some of the information which newcomers need to know, to do their jobs or to act safety and within company rules, is given at the right time — for example, fire prevention needs to be given on the first day, whereas information on the company safety policy is not so urgent.

As there is a limit to the amount of information which new employees can absorb at any one time, induction training should preferably be spread over a period which could be up to three months. Managers and supervisors should be encouraged to continue the process by periodic informal talks with employees dealing with such topics as organizational changes, new products and changes in personnel policy.

Pay special attention to the induction training needs of disabled people; their individual needs should feature in any subsequent training as they change jobs. Consideration should be given to possible modifications to equipment, the use of special employment aids or job restructuring where appropriate, adjustments to the working environment — for example, ramps for wheelchairs, rails for guiding blind people.

It is also important, after getting the employee's consent, to ensure that the supervisor understands the nature of the disability and any implications it may have for the working routine, and to see that fellow employees and trade union and work place representatives are sufficiently informed. They may need

## TABLE 3.1
## Safety induction check list

1.  Company profile
2.  Safety policy
3.  Health and Safety at Work etc Act (general duties — employer and employee)
4.  Other relevant legislation
5.  Training and skills development programme
6.  General philosophy on safe systems of work
7.  Company safety culture and safety management systems
8.  Safety Representative and constituency
9.  Safety meetings and committees
10. Accident and incident reporting requirements
11. Job specific rules and procedures
12. Personal protective equipment — supply and replacement
13. Emergency response procedures
14. Prohibited articles
15. Environmental awareness

### Potential hazards

1.  Basic hazards of operations
2.  Fire and explosion and their prevention
3.  Hazardous substance and safe handling procedures
4.  Confined space operations
5.  Electricity
6.  Crane operations, lifting and slinging
7.  Noise
8.  Scaffolding
9.  Materials handling
10. Housekeeping standards

to agree to modified working arrangements, and there is a possibility that the disabled employee may need help.

## JOB TRAINING

Before the training for any job is carried out, analyse the training needs of that particular job. These needs should be defined to a degree of detail sufficient for the health and safety aspects of the job to be identified and an effective training programme to be prepared.

The definition of training needs starts with a description of the job. For simple jobs a brief list of tasks is sufficient, but for complex jobs a more detailed description may be necessary. The training needs of particular jobs can be broken down into four sections:

- the position of the job in the organization and the related relationships;
- regular features of the job;
- non-regular features of the job, including special hazards and necessary precautions;
- special knowledge and skills required.

The following are examples of sources of information which could be used in describing the job:

- operating instructions;
- job descriptions;
- company policies and procedures;
- analytical or standard methods;
- manuals.

For complicated or hazardous jobs it is best to carry out a job or task analysis in which the job is broken down into a series of discrete operations. Each operation is then further broken down into discrete steps. Each step is analysed to identify the key or critical points, including safety requirements. This allows the skill and knowledge required to be identified (see Table 3.2). It is then possible to identify or construct suitable training sessions to impart the knowledge and practical training sessions to allow the required skills to be built up.

The following items should be considered in the training programme:

- the elements to be learnt;
- time allocation;
- level of performance to be achieved;
- who will arrange the instruction;
- where the training will take place;
- a means of determining that the trainee has achieved the standard which was set.

**TABLE 3.2**

**Job analysis**

| Job title: | | | |
|---|---|---|---|
| No. | Operation | No. | Operation |
| 1. | | 5. | |
| 2. | | 6. | |
| 3. | | 7. | |
| 4. | | 8. | |

**Operation No. 4**

| Tools required: | | | | | |
|---|---|---|---|---|---|
| No. | Step | Key points (including safety requirements) | Skill | Knowledge | Training needed |
| 4.1 | | | | | |
| 4.2 | | | | | |
| 4.3 | | | | | |
| 4.4 | | | | | |
| 4.5 | | | | | |

The role of the employee's supervisor is a vital one when arranging the trainee's programme. As noted earlier, the supervisor's attitude towards training and the staff will determine, to a very great degree, the effectiveness of the training. If any part of the programme has to be delegated, the person actually carrying out the training should have received instruction in appropriate training techniques.

The effectiveness of the training which the trainee is receiving should be continually checked while carrying out the training programme. Is the method used the most appropriate? Is the level right? Should it be modified or speeded up or slowed down? When the training has been completed the trainee, trainer and manager/supervisor should discuss the whole training programme to identify how appropriate it was in answering the training needs which were identified, bearing in mind the following:

- training continues until satisfactory standards are achieved — not necessarily for a set period of time;
- acceptable performance should be judged by the trainee's manager/supervisor and, where possible, to quantified standards which had been agreed;
- objective judgements based on tests, questions, etc, should be preferred to subjective judgements.

The effectiveness of training should be judged by reviewing the content of the training programme against the results of the training given; training programmes must then be revised as necessary.

## SAFETY AWARENESS TRAINING

There are three key elements to achieving commitment to safety:

- getting attention;
- achieving commitment;
- maintaining involvement.

### GETTING ATTENTION

The objective is to overcome complacency and achieve recognition of safety as a personal as well as a company issue. This is traditionally achieved by such means as safety briefings, safety posters and safety meetings. It is probable, however, that these will become routine and staff will then believe that they have 'heard it all before' or be satisfied that they have achieved an adequate level of safety.

To get attention, focus on the personal aspects of safety by:

- reviewing recent incidents and their consequences from an individual's viewpoint;
- using photographs, records or other information relating to real or typical work situations and hazards;
- describing social scenarios to highlight the impact and duration of accidents as they affect individuals, colleagues, family, friends, etc;
- explaining the financial implications for the individual following an accident;
- using scenarios and tests to allow attendees to review their real level of safety awareness.

### ACHIEVING COMMITMENT

Convince personnel that familiarity can lead to accidents to themselves. Particularly with a stable work force, there is a strong possibility that groups have developed short cuts and routines that are not in line with good safety practice or defined procedures. Whilst some of those changes in routine may be definite improvements, others may be unsafe but pursued because 'it's never been a problem'. Similarly the potential risks inherent on a plant become hidden by familiarity with the environment.

It is therefore necessary to increase realization that everyday actions have caused and will continue to cause incidents, and also that accident risks can increase with experience and familiarity. Do this by looking at:

- incidents caused by short cuts, and pictures of potential hazards that have been observed/staged on company premises;
- hazards in individuals' own areas and other areas. Compare findings with either the course leader's or another group's observations to highlight the effect of familiarity.

Additionally it is important to convey the message that others' unsafe actions can jeopardize their own safety and that unsafe practices cannot be ignored. This can be achieved through use of pictures/illustrations, or a short case study.


MAINTAINING INVOLVEMENT

Safety awareness requires increased personal association with safety; *maintenance* of that association, however, requires a continuing programme of involvement, together with supervisory and management commitment. Such a programme might include:

- involvement in safety inspections, on a rotating basis, either in the individual's own or other areas;
- responsibility for safety briefings to contractors and other staff working in particular locations;
- responsibility for identifying and progressing safety improvement projects.

Areas which need to be covered in this programme include, but are not limited to:

- position of people;
- action of people;
- personal protective clothing and equipment;
- tools and equipment;
- procedures;
- housekeeping;
- maintenance;
- why accidents happen;
- emergency plans;
- supervisor safety training.

**TABLE 3.3**
**Supervisor safety training topics**

| | |
|---|---|
| Understanding safety | • safety awareness<br>• attitude and motivation<br>• safety promotion<br>• recognition schemes |
| Responsibilities | • company safety policy<br>• individual's responsibilities<br>• supervisor's responsibilites<br>• company/management responsibilities |
| Legislation/standards | • statutory instruments<br>• codes of practice |
| Hazard identification | • unsafe acts and conditions<br>• cause and effect<br>• corrective actions<br>• inspection of the work place |
| Accident prevention | • definitions<br>• who is responsible?<br>• safety meetings<br>• drills and exercises<br>• training on and off the job |
| Incident and accident reports | • what constitutes an accident<br>• what constitutes an incident<br>• near misses<br>• handling an accident<br>• unsafe acts and conditions<br>• basic causes<br>• corrective actions<br>• follow-up |
| Protective and safety equipment | • the right apparel<br>• to wear or not to wear a hard hat, etc<br>• eye protection<br>• respiratory protection<br>• hand and foot protection |
| Summaries and analysis | • categories<br>• lost time or no lost time<br>• determination of cause<br>• analysis, action and elimination |

**TABLE 3.3 (continued)**
**Supervisor safety training topics**

| | |
|---|---|
| Loss control | • personal injuries |
| | • damage to equipment |
| | • material damage |
| | • effects on the environment |
| | • down time |
| | • damage to property |
| | • employee confidence |
| | • humanitarian considerations |
| | • public concern |
| Risk management | • medical costs |
| | • payroll burden |
| | • loss insurance and deductibles |
| | • compensation |
| | • litigation |
| | • uninsured risks |
| Cost effectiveness | • human resources management |
| | • material management |
| | • time management |
| | • awards — are they worth it? |
| | • financial management |
| Auditing the system | • inspections and audits |
| | • safety committees |
| | • analysis of results |
| | • suggestions and follow-up |
| | • the audit system |
| | • pre-use equipment check |
| | • critical parts inspection |
| Job procedures | • critical job inventory |
| | • job analysis |
| | • planned/spot job observation |
| | • plant safety rules |
| Training | • skills inventory |
| | • developing training requirements |
| | • company policy and standards |
| | • orientation |
| | • job training |
| | • safety meetings |

**TABLE 3.3 (continued)**
**Supervisor safety training topics**

| | |
|---|---|
| Safety Representatives and Committees | • Safety Representative elections<br>• Safety Representative activities<br>• Safety Committee organization, terms of reference and activity |
| Hazardous environments | • dangerous substances<br>• noise and vibration<br>• radiation<br>• electricity<br>• pollution<br>• machinery<br>• entry to confined spaces<br>• purchasing<br>• guidelines for safe contractor operations<br>• manual handling operations |
| System of work | • permits-to-work<br>• isolation, lock, tag and try |
| Disaster control | • contingency planning<br>• directory of key personnel<br>• emergency procedures<br>• management responses<br>• external resources |

Supervisors are, broadly speaking, the only level of management which has direct control of the work force rather than indirect control through the medium of other managers. They are the link between more senior management and the 'shop floor' and are, therefore, a fundamental channel of communication in both directions. The effectiveness of supervisors depends on a number of basic conditions:

• the organizational structure of the firm is sound;
• the supervisor's job is clearly defined and worthwhile;
• supervisors have adequate delegated authority and support from senior management;
• supervisors are carefully selected and trained.

As a prerequisite to training for the supervisor's job a number of steps are necessary. The first is to prepare a job description based on experience and observation of the job. This defines the duties, responsibilities and objectives of

28

the job, the 'key' areas and their main elements, the organizational (line of command) relationships with superiors and subordinates (including limits of authority), and working relationships with colleagues in related units or departments with whom co-operation and liaison is necessary.

The next step is to determine for each activity of the job the knowledge and skills needed to do it. For example, a supervisor responsible for on-the-job training of new recruits needs to be proficient in techniques of instruction.

Then compare the knowledge and skills needed with those already possessed by the trainee. From this the training needed by a particular individual can be deduced — for example, skill in interviewing.

The analysis of training needs is itself a skilled operation, and the manager who has the responsibility for it may need appropriate guidance (for example, from a Training Officer). A list of possible safety topics is given in Table 3.3 on pages 26–28.

## THE TRAINING MATRIX

Having generated a list of the skills required it is very useful to formulate the training needed by the various work groups by generating a matrix listing all training modules needed along the top, the work groups down the side and showing who needs what. See Table 3.4 for a sample matrix.

## TABLE 3.4
## Training matrix

| Work group | Title | General safety | Systems of work | Permits | Isolation | Confined space | Scaffolding | | | |
|---|---|---|---|---|---|---|---|---|---|---|
| Production operators | | | | | | | | | | |
| Maintenance electricians | | | | | | | | | | |
| Maintenance mechanics | | | | | | | | | | |
| | | | | | | | | | | |

# 4. WORK FORCE INVOLVEMENT AND MOTIVATION

The provision of a safe place of work and the development of safe systems of work will not in themselves guarantee safe operations. After all, as the saying goes, 'you can take a horse to water but you cannot make it drink'. So safety can only be achieved if the whole work force actually uses the facilities, systems and procedures correctly. In fact it can be argued that an unsafe place of work need not cause an accident if the personnel working in it are aware of the hazards at all times and ensure that they do not allow the dangers to translate into occurrences. This must not be taken to mean that strenuous efforts to provide safe facilities should be abandoned. They are necessary because the work force is composed of human beings who are fallible, and in addition it is morally and economically wrong to make life difficult. It is much better to do everything reasonable to make it easy to do the safe correct thing.

What this discussion is trying to show, however, is that no matter how good the equipment and procedures are, it is absolutely essential that everyone is motivated to work safely. In addition, as is pointed out in Chapter 2, the best source of detailed knowledge about how a process or procedure actually works in practice is the people who do the job. This also applies to safety and hazards — thus the work force should not only be motivated but also involved in the safety effort and consulted on how things can be done safely. This has the advantage, in addition to providing practical knowledge, of developing ownership and hence application and involvement. These two principles of safety — motivation and involvement — help to create the 'safety culture' mentioned in Chapter 1. To help sustain this culture, recognition needs to be given to encourage people to continue their efforts and try even harder.

## CONSULTATION AND INVOLVEMENT POLICY

A work force consultation and involvement policy needs to be developed and publicized. This policy clearly sets out the company aim of involving everyone in the safety effort, and describes how consultation with the appropriate groups on the provision of facilities and a safe place of work, development of safe systems of work and the implementation of relevant safety programmes and activities will be carried out. The policy should also make it plain that the company will not tolerate any action, real or implied, which tries to intimidate or victimize anyone because they raise safety concerns. Finally the policy needs to spell out how anyone, employee or contractor, who believes that they are being intimidated or victimized can appeal to the highest levels in the company. Appendix 1 gives a possible policy for consideration.

The policy, once endorsed by the Chief Executive, should be issued to all new starters during their induction/orientation programme and posted in all work places owned or under contract to the company. In addition, before any new contract starts, a meeting should be held with the contract company management (as mentioned in Chapter 9) and the work force consultation and involvement policy discussed. It is best to try to get the contract company to issue a copy of the policy to all its employees who will be involved on the contract.

## WORK FORCE ATTITUDES

Common themes which are apparent in a number of surveys of work force attitudes in the offshore oil industry and which may also apply to other industries are:

• industry is adopting a more open system of communication with its work force and encouraging the internal flow of information, including contractor personnel. This process is to be encouraged;

• it is necessary to reinforce continually, and to demonstrate by physical actions, that nothing is more important than safe operations;

• whilst employees generally feel no restrictions or inhibitions on raising and discussing safety concerns within their own company, many contractor employees are reluctant to raise these matters with the client company supervisors, despite the provision of anti-victimization policies. A wider recognition and reinforcement of these policies may help to overcome this reluctance;

• Safety Representatives are a valuable resource in the efforts to attain safe operations. Nevertheless it is necessary that their role is properly defined. It is essential that they do not interfere with or usurp the role of the supervisor; equally they are not safety policemen or safety advisors. Their role and responsibilities need to be clearly defined and agreed. This will help to encourage personnel to come forward for nomination as Safety Representatives.

## SAFETY REPRESENTATIVES

Safety Representatives can play a very important role in the safety programme. Provided the correct 'culture' is in place they provide a useful focus for discussions on possible changes and developments, since they will input valuable information on work force attitudes and practical knowledge. It is therefore recommended that companies develop a system of voluntary Safety Representatives where there are no statutory Safety Representatives. Where union representatives cover only part of the work force, efforts should be made to get them to accept voluntary representatives as representing the rest of the work force.

Safety Representatives must be properly trained. As soon as possible after their election they should be released for a union training course or, if voluntary representatives, given an in-house course covering:

- relevant safety legislation;
- communication and interpersonal skills, written and verbal;
- consultation with constituents, management and the regulatory authority;
- function and powers;
- Safety Committee function and conduct;
- accident prevention;
- safe place of work;
- safe systems of work;
- safety inspections;
- accident and incident investigation;
- sources of information.

Safety Representatives should be given refresher courses after a year in office to brush up their knowledge and skills.

In addition to the functions defined in legislation, Safety Representatives can contribute significantly to the safety effort by complementing the company safety management system. It is essential that the Safety Representatives are not seen as a separate entity but as an integral part of the safety structure. It is suggested that the Safety Representatives could assist in some or all of the following activities in support of the current safety programme and activities:

- Hazop studies as appropriate;
- field safety audits;
- safety inspections and housekeeping inspections;
- accident investigation and analyses;
- safety training;
- emergency exercise training;
- accident reviews;
- safety brainstorming sessions;
- safe system of work audits;
- toolbox safety sessions.

## SAFETY COMMITTEES

To co-ordinate activities and maximize input from employees, the company should develop a Safety Committee structure. Depending on the size and complexity of the company, there may be one Safety Committee or a series of Plant Committees feeding into the Works Committee which in turn reports to the Company Committee. Large corporations require an even larger committee structure.

The top committee should be chaired by the Chief Executive and be composed of the senior managers plus representatives from each subsidiary committee. The objective of this committee is to monitor safety, health and environmental performance throughout the company, review trends, problem areas, accidents and incidents, involve employees in the planning and running of safety programmes and activities and recommend new safety policies, initiatives and programmes. The committee should meet at least quarterly. The Plant Committee should be chaired by the Plant Manager and composed of the Safety Representatives and relevant supervisors. This committee should meet monthly and have the following objectives:

- review actions arising from the previous minutes;
- review relevant safety inspections, discuss hazards and recommend solutions;
- discuss any accidents and relevant incidents;
- review company performance;
- consider suggestions and ways of improving safety performance;
- nominate a representative to attend the higher committee.


## SAFETY MEETINGS

Safety meetings should be held in all facilities so that every employee and contractor employee can attend a meeting. The meeting should be chaired by the relevant supervisor. The frequency varies with the level of risk — say, monthly in process facilities and quarterly in office facilities. The objective is to foster and encourage safety awareness and participation by everyone in the safety effort. The agenda should:

- review the implementation of agreed improvements;
- discuss hazards and safety problems;
- review changes in policy;
- include the presentation and discussion of a relevant safety topic.


## MOTIVATION

To understand why people have accidents and do unsafe things requires a knowledge of human behaviour. We are surrounded by unsafe situations and

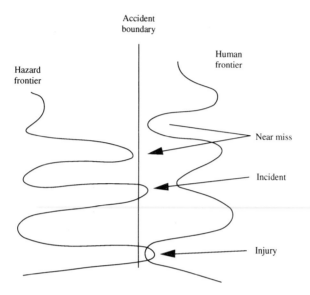

Figure 4.1    Human and hazard frontiers.

hazards at all times, yet we do not have accidents continuously. In fact, accidents are quite rare events. There appears to be a moving frontier of hazard which can on occasions cross the accident threshold and an incident occurs. Equally there seems to be a human behaviour frontier. When this crosses the hazard frontier an injury occurs (see Figure 4.1). However, this still does not explain why it varies, and why some people have accidents whilst others in the same situation do not.

Two theories have been put forward for this state of affairs:
- the accident-prone or careless worker;
- work itself is dangerous to your safety.

As accidents are not everyday occurrences for most people and happen in situations which have occurred many times before without mishap, it follows that accidents are abnormal. It is only a small step to think that people involved in accidents are also abnormal. If it is not normal for most people, it must be the individual who is to blame. Nobody of course suggests that all, or indeed most, accidents are due to psychological or social factors. Nevertheless the evidence points to certain personality and emotional factors as playing an important part.

So why do people allow themselves to get into situations where they have an accident? A.H. Maslow, in his studies on motivation (see *Motivation and Personality*, 1954, published by Harper), found that we all have needs that we seek to satisfy. Figure 4.2 shows these needs in their order of importance to us. It indicates that for most of us health, comfort and money come pretty low in our list of what really drives us. It can be seen that the highest levels of motivation come from achievement and creativity. According to Maslow's model, the higher motivators should override the lower ones once the lower motivators have been satisfied to a reasonable extent. Of course it is a matter of relative values. The approval of friends may not be of great value if the person can gain a lot more money; he may be prepared to risk unpopularity if the rewards are high enough.

Similarly people will risk their health and safety to win social approval by conforming to the current norm of behaviour and not wear protective equipment. They are even more likely to do so to earn status, self-esteem or a sense of achievement. Some people risk their family relationships and ignore social pressures in the search for rewards in achievement and self-esteem, yet the motivation to avoid harm to their loved ones would override the desire to conform to social pressures. In fact, though social pressures are recognized as powerful influences, they only take precedence over those lower in the triangle — namely

Figure 4.2    Maslow' hierarchy of needs.

health, safety, money and comfort. For a lower motivator to take precedence over a higher one, the reward for the lower one has to be very great to compensate for the loss of the higher one. Workers usually expect extra danger money for risking their safety, but a man would need a lot more money to compensate for loss of family relationships or self-esteem. The problem with motivating for health and safety is that in most cases the benefits of taking the necessary precautions are very long term, of low motivating value and the risk/benefit ratio uncertain anyway.

When making a decision a person must weigh up the 'expectancy' of the situation. Take a male process operator, for instance. He will have to put a value in his own mind on the likelihood or probability of various events happening — for example, the odds on him having an accident, the odds that his mates will think he is a cissy if he wears protective gear and the odds that the gear will be effective. His assessment of his chance of avoiding the accident based on his perception of his skill and ability, and the frequency of such accidents, will both contribute to his decision about whether to use the safety gear or not. If he rightly perceives the risk to be slight and the severity of any resulting accident to be minor, it is a waste of time trying to persuade him to take uncomfortable or unpopular precautions. His losses in discomfort outweigh his gains from increased safety. Where the perceived risk is high, and the severity of the consequences great, there is likely to be little problem in persuading people to take precautions. Unfortunately most situations that face us lie somewhere in between.

The last point leads naturally into a description of an important theory of behaviour change called cognitive dissonance. This process of decision-making results over a period of time in a routine or habitual type of behaviour which causes little bother to the individual. When new information arrives or a new decision is required, a state of 'dissonance' is set up. This state of psychological tension is unbearable to most people and they therefore make an effort to change it. This applies only to beliefs or behaviour which are in conflict or inconsistent, not just to 'for and against' arguments.

For example, a man who believes it is a waste of time to wear eye protection and who goes to a new job where it is accepted practice to wear has a problem. He finds that he is ostracized by his new mates unless he wears eye protection. He now has two conflicting beliefs: 'wearing eye protection is a waste of time'; and 'I shall be unpopular if I don't wear it'. He has three ways of resolving the dissonance:

- he can change his behaviour and wear the eye protection. To reduce his dissonance he will need to persuade himself that wearing eye protection is not such a bad thing after all, by thinking up extra reasons in favour of it, or believing that his arguments against it were not important;

- he can refuse to wear the eye protection and use strong counter arguments to persuade the rest of the crew to do the same;
- if this is not successful, he will need to persuade himself that he does not care what the rest of the group think or leave the group.

The first alternative is obviously the most satisfactory, and for this reason it is often found that if behaviour is changed first, a change in attitude can follow. Thus by changing the ratio of consistent and inconsistent beliefs, he returns to a relaxed state and the tension is reduced. This process is called 'rationalization' and because beliefs and values are constantly challenged, this technique is used on many occasions. When a keen disappointment is sustained there is often a period of 'coming to terms with it', which involves self-persuasion that the previous expectations and values were wrong and don't matter as much as was thought. Reasons for accepting the situation are sought in order to reduce the psychological tension. If the consequences of the decision made are rewarding, it is more likely that the same decision will be made again. The beliefs, values and probabilities are 'reinforced' by the feedback. If the consequences are unpleasant (either physically, emotionally or socially), the beliefs, values and judgement of probabilities are altered and next time a different decision will be made. This method of learning by punishment and reward, trial and error, is taking place unconsciously all the time. When applied by managers it is called 'behaviour modification'.

This method can be very successful, but it is usually found that reward is more effective than punishment. This process is illustrated in Figure 4.3 on page 38.

## PHYSIOLOGY

Things are done by the body because information is received by the senses and transmitted to the brain, where it is compared with expectations and memory to reach a decision. The decision is then transmitted to the muscles for conversion into action. Anything which affects this process influences the resultant action — for example:

- senses — design of display, physiological deficiencies;
- expectations — switch down for on in UK and up in USA, estimation of risk;
- memory — short-term memory holds five to ten items, limitation on process rate and capacity;
- muscles — physical dimensions and abilities, reaction time, physiological deficiencies.

People follow certain stereotypes — for example, rotate things clockwise to increase something. They can be taught to operate against a stereotype,

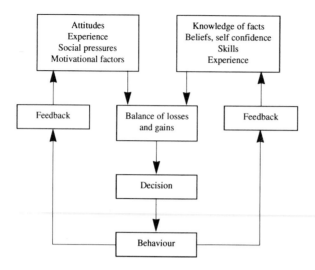

Figure 4.3    Factors affecting behavioural decisions.

but at a certain level of difficulty or stress they reach their limit and any increase in stress causes them to revert to the stereotype. Thus it is necessary to ensure that the information received and the actions required are in line with the stereotype. This all builds up into the picture depicted in Figure 4.4.

Recent work by William A. Wagenaar of Leiden University and James T. Reason of Manchester University (paper 23248 presented at the Society of Petroleum Engineers (SPE) conference in The Hague, 1991) suggests that there are ten areas of latent failure in work situations:

• hardware;
• design;
• maintenance;
• procedures;
• housekeeping;
• error-reinforcing conditions;
• incompatible goals;
• training;
• communications;
• organization.

Wagenaar and Reason disagree with Maslow and assess the effectiveness of various methods of inducing safe behaviour as shown in Table 4.1.

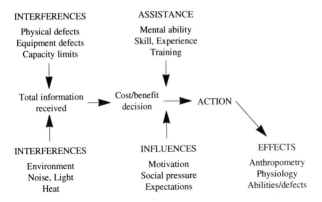

Figure 4.4    Psychology and physiology of behaviour.

## TABLE 4.1
### The assessment of behaviour-influencing methods

| Action | Cost | Effect | Assessment |
|---|---|---|---|
| Make the system foolproof | High | Low | Poor |
| Tell them what to do | Low | Low | Medium |
| Reward and punishment | Medium | Medium | Medium |
| Raise motivation and awareness | Medium | Low | Poor |
| Select personnel | High | Medium | Poor |
| Change the environment | High | High | Medium |

Building these six behaviour-influencing methods into the general accident causation scenario gives the picture shown in Figure 4.5 on page 40.

### SAFETY CULTURE
Whichever theory is believed, it is essential that the correct culture is present in the company. A 1990 UK Confederation of British Industry report entitled *Developing a Safety Culture* identifies the following features as essential:
- leadership and commitment from top management which is both genuine and visible;

- acceptance that the strategy requires sustained effort and interest;
- a policy statement with high expectations and optimism;
- adequate standards and codes of practice;
- health and safety equal in importance to all other corporate aims;
- line management responsible for safety and health;
- ownership of health and safety must permeate all levels of the work force;
- everyone must be involved;
- everyone needs training;
- open communication throughout the organization;
- realistic targets must be set;
- performance must be measured against the targets;
- incidents and near misses must be investigated;
- behaviour must be audited against agreed standards;
- safe behaviour should be a condition of employment;
- identified deficiencies must be remedied promptly;
- management must be given adequate up-to-date information to assess performance.

An interesting discussion by Alan Waring on how to develop a safety culture was published in the *Safety and Health Practitioner* in April 1992. This article discusses the complexities of a company's culture and the problems involved in changing it.

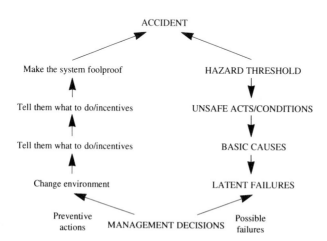

Figure 4.5    General accident causation model.

## RECOGNITION

Maslow's hierarchy of needs puts financial incentives well down the order of priority. Equally F. Herzberg, in his study on the *Motivation to Work* (published by Wiley, 1959) suggests that financial gain is a hygiene factor and not a motivator, thus having only a limited effect and operating for a fairly short time span. On this basis safety incentive schemes are not likely to be an effective use of limited finances. It is possible that short-term gains will be achieved but are unlikely to be long-lived, as the financial benefits soon become the norm and in fact the incentive frequently changes from operating safely to hiding the incidents and making them appear as less serious. It is possibly more effective to use financial incentive to meet needs higher up Maslow's hierarchy than in take-home pay. The finances can be used to generate incentives in the fields of self-esteem, by donating money to charity (especially a local one), or by enabling an achievement to be reached by adding to employees' efforts in a local good cause — for example, high tech equipment for the local hospital, a guide dog for a local blind person. To reinforce the safety incentive, the safety performance recognition programme should reward effort and good performance. The award should be a suitable gift to employees — such as a smoke detector, car safety equipment or fire extinguisher — plus a donation to a local charity or charitable fund-raising effort. The basis for the award should be the achievement of significant safety milestones. In addition, individuals, employees and contractors should be awarded items of token value but publicly recognizable — for example, a sports bag or bomber jacket with a recognizable design or logo — for innovative safety suggestions or significant safety awareness.

# 5.    A SAFE PLACE OF WORK

Safe operation requires a safe place of work. To ensure a safe place of work facilities must be designed, constructed and maintained to an acceptable standard. A safe place of work has the following characteristics:

- safe access and egress for normal and emergency conditions;
- structures and buildings which will withstand normal and abnormal environmental conditions, earthquakes and other natural and man-made hazards (for example, flooding or crashing aircraft) which can be reasonably expected;
- facilities and equipment which can be started up, operated and shut down safely without risks to health or the environment and have a very high integrity of containment and limitation of inventory;
- systems which allow the plant to be shut down, vented/blown down and isolated safely in an emergency, and which are automatic in action where possible;
- facilities and equipment which will contain and control any emergency which can reasonably be expected to occur, so minimizing the risk to personnel, property and the environment both on and off company property.

To achieve this facilities must be:

- designed to accepted design standards or good industrial practice;
- reviewed for hazards to personnel, property or the environment at the feasibility, concept and detailed engineering phases using appropriate techniques;
- audited prior to start-up to ensure that the facilities, software and personnel are ready and able to start up and operate safely;
- inspected and audited throughout their operating life to ensure that adequate standards of safety are being maintained.

Figure 5.1 shows the various steps of a review to ensure that a safe place of work is provided. First, define the objectives and acceptance criteria. Then identify all possible accident events and assess the hazard from each, along with the level of risk. If the risk level is acceptable, the scenario becomes a residual accident event and it will be necessary to generate an emergency response to identify how it will be coped with if it does occur. If the risk level is unacceptable the scenario is a design accident event and must be eliminated, reduced to an acceptable level or a safe system of work must be developed which will allow the job to be done safely.

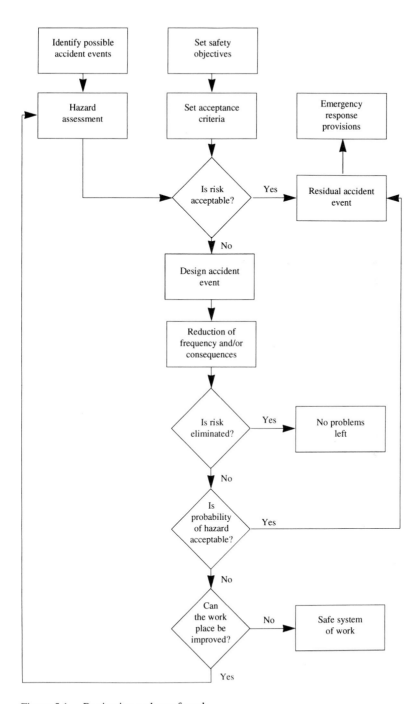

Figure 5.1    Reviewing a place of work.

Modification to existing facilities must also be designed to current good industrial practice, or at least to the same standards as the original design, and must undergo an appropriate safety review.

Existing facilities must be maintained to at least the original design standard, tested and inspected as necessary to detect any deterioration in equipment or facilities and audited at least every five years to confirm that the standards are still acceptable.

Of course the objective is to design a facility which is inherently safe. The October 1987 *Safety and Health Practitioner* article by J.L. Hawksley entitled 'Risk assessment and project development' includes an interesting discussion on this subject. The broad principles for greater inherent safety are to:

- avoid processing or using toxic, flammable or other hazardous materials;
- reduce inventories of hazardous materials to the very minimum;
- reduce the potential for surprise;
- separate people from hazards.

These principles can be applied systematically by considering every aspect of the process, plant and equipment using the procedure given in Figure 5.2.

## FEASIBILITY STAGE

Every project proposal should include a section which shows that the project can be completed and operated without violating any legislative requirements or company policy or standard. This will require the assessment of potential hazards and possible safety and environmental impacts. It is essential that it be clearly shown that acceptable solutions are available. This information must be incorporated in the basis of design for the project.

## CONCEPT STAGE

Once the company has decided that the proposal fits into its long-term plans the project moves into the concept stage in which the various options for the design of the project are evaluated. Each option is developed to the stage at which it can be properly evaluated from the following points of view:

- technical;
- economics;
- operability and maintainability;
- safety and environmental aspects.

At the end of the concept stage a preferred option emerges and a full report describing the options, their attributes and problem areas is available for endorsement by senior management. The preferred option description must contain a safety plan based on qualitative safety and environmental studies including, where appropriate, a 'coarse Hazop'.

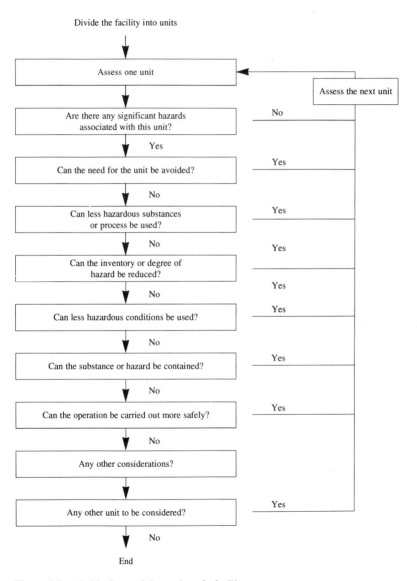

Figure 5.2    Achieving an inherently safe facility.

## DETAILED ENGINEERING

One of the first actions following approval of the preferred option is the definition of the scope of the project. This is best included as part of a basic data manual which is used by the Project Team in preparing the design of the facility and the Operations Team in planning for start-up and operations. The manual should include:

- the overall scope of the project, with a definition of what is included and what is excluded from the project;
- the storage facilities, utilities and ancillary facilities to be provided as part of the project;
- materials involved, including raw materials, intermediates and finished products giving specifications, chemical and physical properties and safety and environmental hazards and precautions;
- the process flow sheet, including mass and heat balances;
- the basic engineering flow sheet along with the basic concept of how the facilities are to be operated;
- the standards to be used in the design;
- the safety and environmental data and plan to be used in the design;
- the coarse Hazop, if carried out.

Experience of a significant number of projects has shown that much remedial and additional work has to be carried out following completion of construction to make the plant operable and to ease plant maintenance. Many of these modifications and additions would be unnecessary and significant cost savings achieved if they had been incorporated into the original design. A better result can be achieved by ensuring early and full involvement of the Operations Team in the design stage, locating them near the Design Team and ensuring regular and full contact and review.

During the detailed design stage, appropriate safety studies must be carried out to allow a full safety plan to be developed. The plan lists all the significant hazards and the controls being applied to contain them. The appropriate industry design codes and standards should be used as yardsticks. In addition to the traditional safety assessments, many companies now carry out reviews of human factors — ergonomics — using suitable specialists. Their advice can contribute significantly to creating a safe and efficient work place.

## OPERATION

Periodic audits should be carried out throughout the operational life of a facility to confirm that the facility is still acceptable and that the operation can still be carried out safely. It is strongly recommended that an audit of the safety control system be carried out about six months after start-up to ensure that the equipment provided is satisfactory and operates correctly.

## MODIFICATION OF EXISTING FACILITIES

Because modification of equipment can have a major impact on the safety of operations and facilities, all modifications should be reviewed and approved by a competent engineer and, where appropriate, by the Safety Department prior to implementation. It is therefore essential that a written procedure be issued detailing how modifications will be reviewed and approved prior to implementation. The engineer working on the project should, in conjunction with the supervisor, decide the level of safety review required. If they decide on the use of a safety check list, the engineer should complete it, review it with the supervisor and file it with the project documentation. If the review involves a mini or full Hazop, the engineer should invite the Safety Department to take part in the review. If the modification warrants a formal safety review, Production and Safety will have to be involved in the review and approvals. The modification review procedure should include providing or updating any necessary operating procedures, drawings and other documentation.

## HAZARD ASSESSMENT

A hazard assessment is an evaluation of the risks to health and safety involved in an operation and followed by decisions on what actions should be taken to remove or control these risks.

If the assessment shows that there is no likelihood of a risk to health or safety, the assessment is complete and no further precautions are needed.

If the assessment shows that further action is required, decide what needs to be done to eliminate or contain the hazard identified. If it is reasonably practicable to do so, take action to prevent anyone from being exposed to any hazardous situation or substance. Where it is not reasonably practicable to prevent people being exposed, the exposure must be controlled. In such cases:

- select the measures to achieve adequate control;
- work out arrangements to make sure those control measures are properly used and maintained; and
- make sure the work force is trained and instructed in the risks and the precautions to take, so that they can work safely.

Unless the assessment is very simple and obvious, it is best to make a written record. Record or attach sufficient information to show why decisions about risks and precautions have been arrived at, and to make it clear to employees, engineers and managers what parts they have to play in the precautions.

If the conditions alter — for example, the introduction of a new process or machine or a change in the substances used — or if there is any reason to suspect that the assessment is no longer correct — for example, an injury or illness occurs — the assessment must be reviewed to take account of these new circumstances.

Apart from very simple assessments it is best to use a multi-disciplinary team to carry out risk assessment. The team should consist of the design engineer/team members, operations and specialist safety engineers and, where appropriate, maintenance personnel. The composite team possesses a capability greater than the sum of the individuals due to the synergistic effects generated.

### SELECTION OF THE REVIEW TECHNIQUE

The decision tree shown in Figure 5.3 provides guidance on selection of the appropriate review method based on various process and risk criteria. Factors involved include the hazard severity, hazard type and process complexity. To use the decision tree the personnel involved must be familiar with the process equipment,

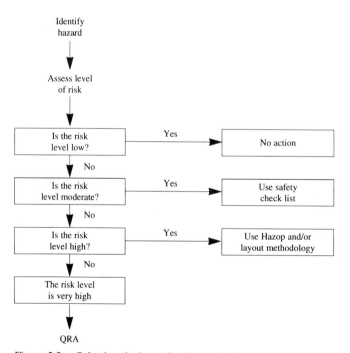

Figure 5.3    Selecting the hazard review technique.

process hazards and operating procedures. They should divide the process into major steps or operations and list all significant hazardous events which could occur with each step or operation. Examples are toxic release, gaseous explosion and fire. The decision tree then provides guidance on the most suitable analysis technique for each hazardous event. The following definitions should be used in conjunction with the decision tree:

- hazard — the ability of something to injure people, property or the environment;
- frequency/probability — the likelihood of an event occurring. It may be expressed as frequency (number of events per unit time) or probability (the chance that the event will occur in any given period);
- population — the number of people (employees, other workers, the public) or units exposed to the hazard;
- exposure — population × frequency;
- consequences — the severity of injury or damage most likely to occur due to the hazard;
- risk — exposure × consequences.

Rate the population and frequency or probability of occurrence as low, medium or high and hence rate the exposure by multiplying the factors together or using Table 5.1. Then rate the consequence as low, medium or high and multiply the exposure factor and consequence rating together to give the risk level. Alternatively use the matrix shown in Table 5.2 overleaf.

## TABLE 5.1
**Exposure matrix**

| Frequency | High<br>1 | Medium<br>2 | Low<br>3 |
|---|---|---|---|
| **Population exposed** | | | |
| High<br>3 | Very high<br>9 | High<br>6 | Medium<br>3 |
| Medium<br>2 | High<br>6 | Medium<br>4 | Low<br>2 |
| Low<br>1 | Medium<br>3 | Low<br>2 | Very low<br>1 |

## TABLE 5.2
## Consequences matrix

| Exposure | High 3 | Medium 2 | Low 1 |
|---|---|---|---|
| Very high 9 | 27 | 18 | 9 |
| High 6 | 18 | 12 | 6 |
| Medium 4 | 12 | 8 | 4 |
| Medium 3 | 9 | 6 | 3 |
| Low 2 | 6 | 4 | 2 |
| Very low 1 | 3 | 2 | 1 |
| Very high risk: 27 High risk: 9–18 Medium risk: 3–8 Low risk: 1 or 2 | | | |

## HAZARD IDENTIFICATION

FIRE AND EXPLOSION HAZARDS
A fire and explosion assessment has three objectives:
- to consider systematically the potential fires and explosions which can occur and to identify those which may cause harm to people. Based on this, judgements can be made on the adequacy of fire and gas detection systems and fire protection systems;
- to provide a structured analysis of the fire and explosive hazards to form a fundamental and integral part of the safety review;
- to identify major deficiencies or weaknesses in the abilities of the protection systems to counteract the assessed fires or withstand the assessed explosions.

It is not appropriate to make recommendations for upgrades or improvements to existing facilities based on the assessment alone. This can only be done satisfactorily as part of an overall review which balances the contribution of these hazards with any other hazards which have a potential to affect people. For example, hardware changes may not be required in all cases, as it may be acceptable to use procedural or management controls to overcome limitations in fixed systems and thus maintain an acceptable level of safety.

This assessment involves the following basic steps:
- define location and the limits;
- collect information on equipment, layout and so on;
- identify inventory composition, pressures and temperatures;
- analyse sources and sizes of all releases;
- assess explosion overpressure, type of fire, duration, heat and smoke emission and so on;
- review possible escalations;
- analyse the effect explosion, heat and/or smoke will have on the facilities, personnel, property and emergency equipment;
- assess the suitability and capabilities of the active and passive fire protection and any other mitigating facilities to fulfil their function.

TOXIC HAZARDS

Information about the hazardous properties of any substance, and its potential for creating risk when used at work, should be included in the basic data manual for any project.

The designer, manufacturer, supplier and importer of any substance used at work may have specific duties under safety at work legislation to provide adequate information. Safety data sheets are a familiar example of ways in which this information can be provided, and are an important source of information for those undertaking substance assessments.

In the UK, any dangerous substances supplied from within the UK mainland should be classified and labelled, in accordance with the Chemicals (Hazard Information and Packaging) Regulations 1993, the Road Traffic (Carriage of Dangerous Substances in Packages, etc) or Harbour Areas Regulations 1987. The hazard information to be provided on the label for certain dangerous substances, including preparations and other mixtures, can be found in the Health and Safety Executive's (HSE's) Approved Carriage List and Approved Supply List. Dangerous preparations and substances that are not on the list still have to be classified and labelled using the principles set out in the Approved Code of Practice *Classification and Labelling of Substances Dangerous for Supply.*

The Control of Substances Hazardous to Health (COSHH) Regulations list specifies substances which have been allocated Maximum Exposure

Limits (MELs). The list is reproduced in Health and Safety Executive (HSE) Guidance Note EH40 which is published annually, and which also lists those substances for which an Occupational Exposure Standard (OES) has been set. HSE publications are available from HSE Books and Dillons bookshops.

It must be appreciated that risks from substances generally arise from the way in which those substances are put to use, and not just from their intrinsic properties, or hazards. The intention should therefore be to identify and assess all work activities which involve, or potentially involve, exposure to substances hazardous to health.

It is not sufficient to assess the work in isolation from the people involved; one task may be performed in different ways by different individuals, and differing practices may be adopted between shift crews.

The assessment must take account of the non-routine as well as the routine. This means that maintenance and emergency procedures need to be subjected to assessment, as well as exposures to materials arising from spillages and leaks.

Having established a basic inventory of substances, together with the relevant work activities, the type and degree of exposure of individuals to the hazardous substances needs to be addressed. All possible routes of exposure should be considered, including inhalation, swallowing, absorption through the skin or eyes and injection by sharp objects or high pressure.

The basic question to be asked is, 'how much are people exposed to, and for how long?' This is best answered by bringing together the experience and knowledge of the people who actually carry out the work in question. In more complex cases the help of a professional occupational hygienist might be required, but frequently the necessary assessment can be carried out using readily available information, applied with common sense.

This involves:
- defining the location and the limits;
- collecting information on equipment, layout, operation and procedures;
- identifying inventory composition, pressure, temperatures and so on;
- analysing sources of release;
- assessing release quantities, concentrations and so on;
- analysing effects on personnel.


## SAFETY ASSESSMENT TECHNIQUES

Safety assessment techniques vary from relatively simple reviews carried out by a single person to very complex and detailed multi-disciplinary team reviews. In every case it is essential to ensure that the reviewers are properly qualified and adequately experienced in the technique to be used and the technology of the subject being reviewed. The techniques include:

- review using a check list;
- layout review;
- Hazop;
- quantified risk assessment (QRA).

## CHECK LIST

The safety check list technique is based on the 'what if' approach in which the process is reviewed from start to finish asking 'what if' at each step to evaluate the affect of any component failure or procedural error on the process. This is best organized using a list of words or phrases to stimulate questions concerning the subject. It is necessary to develop a check list specific to the type of operation and process; a variety of check lists have been published and can be useful as starting points. Table 5.3 on page 54 presents a basic check list for a chemical plant. The *Guide to Engineering Safety Reviews and Audits for Process Plant Contractors* published by the UK Energy Industries Council contains further examples.

## LAYOUT METHODOLOGY

Shell International in The Hague have developed a methodology to improve the design/review of plant layouts. The methodology was described in paper 23269 at the Society of Petroleum Engineers (SPE) conference in The Hague in November 1991 by K.W. Waterfall, J.D. Stockley and P.D. Bentley. The procedure involves a hierarchical approach to layout development. The relative locations of fundamental functions are established using a notional shape taking due account of the project constraints. Then the inherent active and reactive behaviour characteristics of the various functions and individual pieces of equipment are used to confirm layout preferences both in relative location and separation requirements. Finally the scales of potential incidents are considered to determine whether any of the accidental events could reach adjacent areas of the facility and cause serious escalation in the absence of adequate separation or the provision of barriers. Equipment reliability and accidental event frequency are used to determine the frequency of occurrence to assess the level of risk. In addition to safety, operability and maintainability can be considered.

## PURE HAZOP

The need to check designs for errors and omissions has been recognized for a long time, but has traditionally been done by individuals. Experts have usually applied their special skills or experience to check particular aspects of design.

**TABLE 5.3**
**Basic safety check list for a chemical plant**

| | |
|---|---|
| • inventory | • flash arrestors |
| • temperature | • fire and gas detection |
| • pressure | • fire protection and fighting |
| • flow | • explosion prevention and relief |
| • level | • means of escape |
| • composition | • area classification |
| • toxicity, flammability, radioactivity | • fail safe |
| • flash point | • personnel safety equipment |
| • reaction conditions | • environmental conditions |
| • transport and storage | • effluents — solid, liqueous or gaseous |
| • start-up | • noise and vibration |
| • shutdown | • static electricity |
| • abnormal conditions | • material of construction and corrosion |
| • emergency shutdown | • pipe and vessel supports |
| • preparation for maintenance | • heating and ventilation system |
| • maintenance procedures | • drainage |
| • inspection and testing | • emergency power and lighting |
| • relief and blowdown | • lightning |
| • separation and segregation | • traffic |
| • proximity to housing | • ergonomics |

For example, the instrument engineer would check the control systems and, having reassured himself that the systems were satisfactory, would put his mark of approval on the design and pass it to the next 'expert'. This kind of individual checking, provided it is carried out conscientiously, obviously improves the design but clearly has little chance of detecting hazards caused by interactions. These hazards are likely to result from the unexpected interaction of seemingly safe components or methods of operation under exceptional conditions. If it is wished to study such interactions in new designs, the combined skills of a group of experts is required. Their total knowledge and informed imaginations can be used to anticipate whether the plant will operate as intended under all possible circumstances.

Essentially a Hazop study takes a full description of the process and systematically questions every part of it to discover how deviations from the intention of the design can occur and decides whether these deviations can give rise to hazards.

The questioning is focused in turn on every part of the design. Each part is subjected to a number of questions formulated around a number of guide words which are derived from method study techniques. In effect, the guide words are used to ensure that the questions, which are posed to test the integrity of each part of the design, will explore every conceivable way in which that design could deviate from the design intention. This usually produces a number of theoretical deviations and each deviation is then considered to decide how it could be caused and what would be the consequences.

Some of the causes may be unrealistic and so the derived consequences are rejected as not meaningful. Some of the consequences may be trivial and are considered no further. However, there may be some deviations with both causes that are conceivable and consequences that are potentially hazardous. These potential hazards are then noted for remedial action.

Having examined one part of the design and recorded any potential hazards associated with it, the study progresses to focus on the next part of the design. The examination is repeated until the whole plant has been studied.

Although the approach as described may appear to generate many hypothetical deviations in a mechanistic way, the success or failure depends on four aspects:

- the accuracy of drawings and other data used as the basis for study;
- the technical skills and insights of the team;
- the ability of the team to use the approach as an aid to their imagination in visualizing deviations;
- the ability of the team to maintain a sense of proportion, particularly when assessing the seriousness of the hazards which are identified.

Because the examination is so systematic and highly structured, it is necessary that those participating use certain terms in a precise and disciplined way. It is therefore essential that the team has been properly trained in the technique and contains an adequate number of members experienced in Hazop.

MODIFIED HAZOP

The pure Hazop method does not always address procedural aspects of operations adequately and it is often necessary to use a modified version of Hazop to include both equipment and procedures analysis. In recent years modified versions of Hazop have been developed for specific uses — for example, drillers Hazop is used to assess the safety of oil well drilling operations. The coarse Hazop is another modified version of the Hazop technique, aimed at identifying the major hazards (including the potential for disastrous interactions between plants) at an early stage in the development of a project. The project must have certain general parameters established to allow the technique to be used:

- material — raw, intermediate, product and effluent;
- unit operations;
- layout.

These general parameters are then considered in turn using a check list of potential hazards. A useful check list for most chemical plants is the following:

- fire;
- explosion;
- detonation;
- toxicity;
- corrosion;
- radiation;
- noise;
- vibration;
- noxious material;
- electrocution;
- asphyxia;
- mechanical failure.

Other hazards can of course be added for particular kinds of process. When the potential hazards are applied in turn to the general parameters any meaningful combination may indicate a major hazard. The procedure can be carried out very quickly by a small group of experienced people.

## QUANTIFIED RISK ASSESSMENT

Quantified risk assessment (QRA) seeks to provide quantified information about risk. It does this by assessing both the frequency with which hazardous events occur and the magnitude of the result of the event when it occurs. The following stages are involved:

- identify the potential hazardous events;
- assess the frequency with which they will occur;
- calculate the likely consequences of the events;
- assess the impact of these consequences;
- calculate the level of risk;
- decide on the significance of the risk levels.

*Hazard identification*
Identify all scenarios which could be hazardous to the facility and estimate their relevance and significance. Rank the relevant scenarios by a qualitative assessment of their potential risk, to identify those which warrant a full QRA.

*Frequency of significant hazards*
Determine the frequency of the initiating events for the various significant hazards from historical data or relevant data sources. It is usually possible to group hazards together on the basis of common initiating events, thus reducing the magnitude of the task.

*Event tree analysis*
Event trees are used to calculate the frequency of each of the possible outcomes that stem from a single initiating event. At each step in the tree the alternative outcomes are shown — thus the tree branches out to each of the many possible ultimate results. The frequency of occurrence of each alternative is used with the frequency of occurrence of the precursor stage to calculate the frequency of each subsequent stage, thus finally giving the frequency of occurrence for each of the possible outcomes. By examining the various branches of the tree it is possible to identify critical steps, the frequency of which may be modified by some suitable mitigating action to reduce the frequency of unacceptable outcomes.

*Failure and effect analysis*
Use of failure and effect analysis is recommended for the analysis of a small segment of a high potential hazard process — such as a reactor or distillation column — in contrast to an entire production operation or an operating building. The purposes of this procedure are to evaluate carefully all aspects of a particular operation and to investigate the potential of each individual and interacting component failure.

This method of analysis may not give adequate emphasis to omissions or errors in operating procedures, incorrect operational sequences in batch operations, or the possibility or probability of operator errors. Additional and separate 'job audit' studies may be needed to evaluate the hazards of these aspects of a process.

*Fault tree analysis*
A fault tree is a graphical representation of the relationships among the components of a system, based on sequences of failure events leading to a single ultimate undesired 'top event'. Logic diagrams are used to portray and analyse the potentially hazardous 'basic causes' and intervening events. A fault tree analysis

should only be carried out by people proficient in the technique. The objective of the analysis is to determine how a serious process incident could occur as a result of failure of individual process components, including the people involved in operations and maintenance and the probability or frequency of failure. It is therefore necessary to be familiar with the process, the processing and control equipment, and operating procedures and practices.

## VERIFICATION

To be certain that the facility is fit for its purpose, some form of Quality Assurance (QA) needs to be employed. This aim is to ensure that the facility is neither overspecified nor underspecified, not complex when simplicity will suffice, not too simple when complexity is essential. The QA system must confirm that the engineers carrying out the project are working in accordance with the agreed procedures and standards. With respect to safety, the QA system must ensure that:

- the range of scenarios and hazards considered is adequate;
- the consequences derived are realistic;
- the failure rates used are taken from an acceptable, recognized data base;
- where engineering judgement is necessary, the engineer making the judgement is adequately qualified and experienced;
- all relevant systems and equipment are included in the assessment.

# 6.    SAFE SYSTEMS OF WORK

A safe place of work is, of course, extremely hard to achieve. Some residual hazards are almost certain to be present and the actual work to be carried out — be it operational or maintenance — inevitably creates more hazards. It is therefore necessary to initiate and enforce a set of safe systems of work. This is often a requirement under statutory law (for example, the UK Health and Safety at Work etc Act 1974, Section 2 (2)) and also a duty under common law. The statutory duty is a strict liability and failure to provide a safe system of work is an offence, while the common law duty is to take reasonable care of employees whilst they are at work. Interestingly, the common law duty is specific to the employer and cannot be delegated, whereas the statutory duty only applies where in practice the employer has the ability to actually control the situation.

But what makes up a safe system of work? There are a number of components, as shown in Figure 6.1.

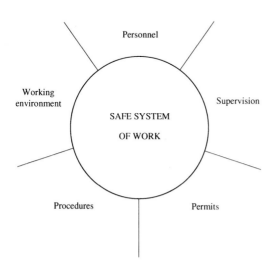

Figure 6.1    A safe system of work.

## PERSONNEL

Personnel must be competent to do the job to ensure a safe system of work. This means that the selection process must be adequate to ensure that qualifications and experience match those required for the job. There is obviously scope for people with lower levels of competence, but they must be provided with suitable training and be appropriately supervised until their competence meets the standard required. There is also a duty to ensure that personnel are physically and mentally capable of doing the job. Asking a person with reduced mobility to do a job which requires agility cannot be considered to be providing a safe system of work.

Having obtained suitable personnel, ensure that they are given an adequate orientation into the company and its systems, and that suitable and appropriate job training is provided.

## SUPERVISION

The provision of adequate supervision is necessary for all safe systems of work. A supervisor is responsible for monitoring adherence to the defined system of work and for motivating personnel. To carry out this monitoring role, a supervisor must know and understand the work being carried out and the systems to be used to ensure safety, and then observe the actual work under way and compare the actual against the standard. Where there is a deviation, corrective action must be taken. The corrective action may be merely a short discussion or may involve retraining of the person involved. In all cases it is necessary for the worker to clearly understand what needs to be done and how it should be done but, equally important, *why* it should be done that way. People cannot be expected to follow the safe system of work automatically — the 'it never happens to me' syndrome applies. Thus it is necessary to motivate people to work safely. Many people play a part in achieving motivation, some being more critical than others, but no-one has a more direct impact than a worker's direct supervisor.

## WORKING ENVIRONMENT

The fixed aspects of the working environment are defined by the safe place of work standards applied to the work place. But many aspects of the working environment are variable and can only be defined once the actual operations are studied whilst under way. Thus whilst the review of the proposed facilities, etc, will involve an assessment of the various hazards listed in Table 6.1 and the provision of suitable mechanical protection, the only way to be sure that these provisions are effective is to monitor actual operations. Where deficiencies are

**TABLE 6.1**
**Work area hazards**

| Hazard | Programme | Safe system of work |
|--------|-----------|---------------------|
| Chemicals | COSHH assessment and industrial hygiene surveys | Appropriate controls |
| Noise | Measurement and personal dosimetry | Appropriate controls |
| Vibration | Measurement | Isolation of area |
| Fire | Risk assessment | Fire prevention |
| Machinery | Review adequacy of guards | Separation of personnel from equipment |
| Physical agent | Assessment and monitoring the agent | Isolation and protection of personnel |
| Ergonomics | Review complaints and sickness records | Redesign work station |
| Strain | Assessment of load and physical effort requirements | Provide mechanical aids and additional labour |
| Moving objects | Assessment of risk | Appropriate controls |

detected, make improvements to the provisions wherever possible. Where this is not feasible, safe systems of work must be provided including any necessary protective equipment.

CHEMICALS SAFETY

It is essential that no substance be brought on site unless a material safety data sheet for it has been received and assessed and a safe handling card generated. The safe handling card is designed to be used by the personnel who transport, store and handle the substance. Thus it must provide easily useable information on the hazards and relevant properties, precautions to be taken to ensure safe handling and storage, first aid requirements and actions to be used in event of a fire or spill. The safe handling card should be displayed at all storage and handling locations.

PERSONAL PROTECTIVE EQUIPMENT

Personal protection includes as appropriate:

- head protection;

61

- facial protection;
- eye protection;
- respiratory protection;
- hand protection;
- arm protection;
- body protection;
- foot protection;
- skin protection;
- fall protection;
- hot/cold protection.

To be sure that the correct personal protection is being used, list all the work on site, including the activities carried out by the operating and maintenance departments and also the contractors who work on site. Check whether it is possible to make any of the jobs safer. Where this is not possible assess the hazards to decide whether personal protective equipment (PPE) is required, ensure that it is of the correct type and gives adequate protection. Useful sources of advice on the correct type of PPE can be found in international and national standards and PPE suppliers may also be able to provide authoritative advice. In the UK, some jobs require equipment approved by the UK Health and Safety Executive (HSE) and from 1994 all PPE in Europe has to carry the European Community 'CE' mark to show that it complies with European Standards. Where there is a choice of PPE available it is always advisable to involve the work force in the selection of the items to be used. Ensure that employees are trained in what PPE they should use, when it is necessary to use it and how to fit and use it correctly. Make sure that an adequate system is in place to ensure that all non-disposable PPE is cleaned after use, inspected and stored in the proper manner.

When carrying out an assessment, consider the consequences of the hazard and the probability of its occurrence. If the combination is high then action must be implemented to contain the hazard or protect the personnel. If the population at risk is high then considerable expense will be justified to contain the hazard, whereas if the population is low and the consequences/probability not high then less expense is justified and personal protection may be adequate.


## PROCEDURES

Well-designed procedures form an essential part of a safe system of work. A procedure defines how a critical task should be done, from the point of view of either safety or efficiency. It allows a supervisor to confirm that things are being done correctly, makes training easier and simplifies the identification of ways

**TABLE 6.2**
**Procedures**

| Type | Coverage |
| --- | --- |
| Operating | Address all steps in each process and all process conditions |
| Maintenance | Address how the facilities will be maintained |
| Specific hazard | Address how specific hazards will be contained and handled safely |

of improving the operation. It is important that procedures relate closely to the actual equipment in place, the instructions provided for it and the people using it. In order to fulfil these requirements, procedures must not be over-complex, and they must avoid gaps, conflicts and inaccuracies. There are various procedures, as shown in Table 6.2, but each of them must explain how to do the job/task safely and, where appropriate, efficiently and to the correct quality. Review procedures at least every two years or when new information becomes available or changes are made to facilities or equipment, and update them as necessary. The use of task analysis as described in Chapter 3 assists greatly in ensuring that procedures are accurate. It helps to identify properly the hazards involved in the operation and the precautions required to ensure safety.

OPERATING PROCEDURES
Operating procedures must be consistent with the process safety information and provide clear step-by-step procedures for:
- normal start-up;
- normal operations;
- normal shutdown;
- emergency shutdown.

There may also be a need for procedures to cover start-up from a major overhaul and temporary operations. The procedures must define safe/acceptable conditions, unsafe/unacceptable conditions and the hazards involved in and precautions required by the operation.

MAINTENANCE PROCEDURES
Maintenance procedures define how equipment will be maintained. They are normally supplied by the equipment vendor but the vendor cannot be aware of the specific operations, hazards and company policies and systems which are used. Because of this it is necessary to review the manufacturer's maintenance

procedures and provide addenda addressing the hazards specific to the operation, the precautions required to contain those hazards and any specific standards of work critical to the activities.

SPECIFIC HAZARDS

Certain jobs are so varied or 'one-off' that it is not practical to write standard procedures, but it is necessary to lay down certain minimum safety standards. Jobs which this applies to include those listed in Table 6.3.

Basically the safety procedure, or local rules for such jobs, should cover the following:
- standard of equipment and protection to be used;
- qualifications and experience of personnel authorized to carry out specific jobs;
- how the work will be controlled;
- where appropriate, the qualifications and experience of personnel who approve the work;
- the inspection system, to ensure maintenance of the standard during the life of the operation or equipment.

**PERMITS**

All non-standard activities which could be hazardous should have a suitable permit issued before commencement of the work. Briefly a permit-to-work (PTW) is a written authorization provided by a Responsible Person certifying that it is safe for others to carry out a specified task in a defined location during a defined time span. The permit also states what steps have been and should be taken to overcome hazards to the persons carrying out the work. Thus a PTW system interacts with the time, the manner and the location to ensure safety, identifying and controlling the people, the hazards and the task.

**TABLE 6.3**
**Specific hazard procedures**

| | |
|---|---|
| • use of portable tools | • use of abrasive blasting |
| • electrical jobs | • high pressure water jetting |
| • radiation handling | • fork-lift truck operations |
| • scaffolding and ladders | • welding |
| • slinging and rigging | |

**TABLE 6.4**
**Activities requiring a permit-to-work**

| | |
|---|---|
| • hot work | • hazardous material handling |
| • cold work | • excavations |
| • confined space entry | • work at high elevations |
| • lethal work | • diving |
| • electrical | • work over water |

A PTW should be raised to cover at least the activities listed in Table 6.4 and be based on the following principles:
• isolation is adequate;
• isolation is secure;
• residual hazards are defined and controlled;
• the equipment to be worked on is clearly and correctly identified;
• the personnel carrying out the task are properly instructed;
• there must be no change in the task without an authorized change to the permit;
• the system is monitored.

Many of these principles will require standards to be defined. The adequacy of isolation must be based on what reasonably foreseeable dangers exist. The level of security of isolation must take into account the potential consequences of failure of the isolation. Where the consequences are serious, a lock, tag and try system should be utilized. This involves the application of a lock to the isolation, and the attachment of a tag to the lock explaining what is isolated, why and by whom. Finally the effectiveness of the isolation is tested by trying to operate the equipment.

It is recommended that a permit should have a limited life — say, 12 hours or until the completion of a shift. If it is desired to extend the life then the work site should be re-inspected by a Responsible Person before the permit is revalidated. Permits should have all relevant certificates attached to them — for example, mechanical isolation, electrical isolation and gas-free test results. They should be signed by a Responsible Person for the area, equipment or system to be worked on, the Task Supervisor responsible for carrying out the work and, where necessary, the Plant Manager to authorize certain types of work and the Responsible Persons for interacting areas. Copies of all permits should be displayed at or near the work site and the plant control centre as well as being held by the Responsible Person. It is essential that there is an effective system

## TABLE 6.5
## Content of permits

| | |
|---|---|
| • work to be undertaken | • any cross-referenced permits |
| • work area | • period of validity and maximum life |
| • equipment to be used | • signature of Responsible Person |
| • identified hazards | • signature of Task Supervisor |
| • precautions already in place | • signature of Plant Manager when required |
| • precautions to be taken | • the handback signatures |
| • all attached certificates | |

for communicating the status of all permits from shift to shift and Responsible Person to Responsible Person.

Many forms of permit are acceptable but they must all clearly show the items listed in Table 6.5.

ISOLATION

Every facility should have a written isolation standard which defines the acceptable standards for mechanical and electrical isolation and the procedures to be followed to install and subsequently remove the isolations. The mechanical isolation standard should include:
• entry to lethal service equipment (if appropriate);
• hazardous service isolation;
• non-hazardous service isolation.
    Methods of isolation include:
• removal of a spool piece and the fitting of blank flanges;
• use of a spectacle blind;
• insertion of a spade;
• double block and bleed;
• locking single isolation valve with two effective sealing surfaces and a vented body cavity;
• single valve.
    Electrical isolation should be by:
• removal of fuses;
• operation and locking of a suitable isolation switch designed to break physically all live and earth conductor paths;
• removal of bus bar links.

Where there is any chance of voltage being induced in the isolated circuit, the system conductors must be earthed.

The isolation procedure should define:

- responsible, authorized and competent persons;
- isolation certificate requirements;
- methods to be used to ensure the continued integrity of the isolation during the work;
- methods to be used to test the effectiveness of the isolation before starting work.

## TRAINING

It is essential that all personnel involved in the PTW system are properly trained, including the Authorizing Authority who approves the permit, the Issuing Authority who prepares the permit and the Performing Authority who carries out the work.

## RISK ASSESSMENT OF THE SYSTEM OF WORK

To put the whole of this aspect of safety together, it is necessary to carry out a risk assessment of the system of work. Look systematically at all the activities carried out on the work site and analyse them to identify all the hazards. Rate their consequences and the probability of occurrence as high, medium, low or nil. The combination of these two gives the level of risk. The measures being used to minimize the risk are then considered to assess the residual risk — see Figure 6.2. The improvements considered should be in the following order:

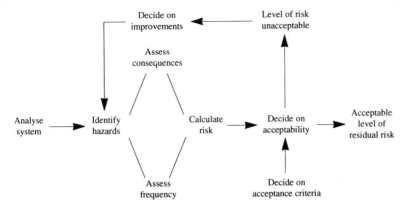

Figure 6.2    Systematical assessment of risk.

- eliminate the risk;
- substitute the risk by a less hazardous activity or substance;
- use engineering methods to control the risk;
- use systems of work to control the risk;
- use protective equipment to protect the personnel.

　　　　Figure 6.3 shows a form that can be used to document the assessment. The final overall assessment should be rated as:

- safe to carry out activity using listed precautions;
- safe to carry out activity using additional precautions;
- unsafe to carry out activity.

| Activity: | | | |
|---|---|---|---|
| Hazard | Effect | Level of consequences | Level of probability |
| | | | |
| | | | |
| | | | |
| | | | |
| | | | |
| | | | |
| | | | |
| | | | |
| | | | |
| | | | |
| | | | |
| | | | |
| | | | |
| | | | |
| | | | |

Figure 6.3　Record of risk assessment.

## MAINTENANCE OF SAFE SYSTEMS OF WORK

Like any other system it is essential to ensure that safe systems of work are reviewed and updated periodically as new hazards are recognized and procedures and operations change, as well as the obvious need to be certain that the agreed systems are actually followed.

Work systems should be inspected at three different levels:

- supervisors continually monitor adherence to the systems of work — for example, procedures and permit requirements — to be sure that the standard of protection specified is being followed;
- supervisors and managers carry out regular inspections daily or weekly to

| Risk | Minimized by | Residual risk |
|------|--------------|---------------|
|      |              |               |
|      |              |               |
|      |              |               |
|      |              |               |
|      |              |               |
|      |              |               |
|      |              |               |
|      |              |               |
|      |              |               |
|      |              |               |
|      |              |               |
|      |              |               |
|      |              |               |
|      |              |               |
|      |              |               |
| Final overall assessment of residual risk | |               |

ensure that the procedures are being followed correctly and that the standard of protection being specified and used is adequate. Work procedures should be reviewed at least annually to ensure that they are still appropriate and describe the optimum way of doing the job. This is best done by the job supervisor and a group of operators fully experienced in the job being studied;

• managers and non-involved personnel audit the PTW system at least annually to ensure that the system and its application is adequate and being followed correctly. Advice on editing and developing permit systems is available in the UK in the Oil Industry Advisory Committee booklet on *Work Permits in the Petroleum Industry* published by the HSE and available from HSE Books or Dillons bookshops. Similar techniques should also be applied to the auditing of procedures and the work environment.

A useful technique for developing and upgrading procedures is the job task analysis. In this, a job is broken down into a series of tasks with each task then subdivided into a series of steps. Each step is analysed for all critical aspects such as safety, quality and efficiency. Then the steps are written up to ensure that the critical aspects are carried out in the manner which is agreed to be the best way to do it. Thus the procedures emphasize critical parameters; don't let them become overly complicated or detailed or 'teach granny to suck eggs'.

Another useful technique to help personnel in following the agreed procedures is to introduce check lists for complicated procedures. These are used in the aviation industry to excellent effect by pilots — for example, prior to take off — but can be applied to any operation.

# 7.   OCCUPATIONAL HYGIENE

A company must take all reasonably practicable steps to ensure that all places of work under its control are without risk to the health of all personnel who could be present — employees, contractors, visitors and members of the public. It has a responsibility to ensure that all places of work are designed, built, operated and maintained to minimize risks to health. Where practical, hazards should be eliminated or at least controlled by engineering means; where this is not practical, personal protection must be supplied. Wherever possible the work force should be involved in evaluating and selecting the equipment from the range which meets the technical specification required to provide the necessary protection. Training in the use of the protective equipment should be provided. Supervisors have a responsibility to audit regularly that the equipment is still functional and clean and that personnel are using it correctly and when necessary. The UK Workplace (Health, Safety and Welfare) Regulations 1992 provide more detailed requirements for many aspects of work place safety.

## HEATING, VENTILATION AND AIR CONDITIONING
Heating, ventilation and air conditioning (HVAC) are provided to:
- supply fresh air and provide temperature and humidity control;
- provide a safety barrier — a pressure differential — against the ingress of flammable or toxic gases or dust;
- dilute and remove small quantities of contaminants such as harmful gases, dusts and odours;
- aid dissipation of heat from equipment.

Care must be taken to ensure that the HVAC function is not impaired by the blocking of grilles, the introduction of new hazards or the defeating of the system — for example, by wedging doors.

Air conditioning systems are a potential source of Legionnaire's disease. Where this is a possibility it is essential that the required testing, treating and maintenance is carried out. Advice on this subject can be found in the following documents:
- HSE Guidance Note EH 22 — *Ventilation of the Workplace*;
- HSE *Principles of Local Exhaust Ventilation*.

## LIGHTING

Lighting systems should be designed to provide adequate illumination for the type of work being carried out. Guidance on appropriate standards of illumination are given in HSE HS(G)38 *Lighting at Work* and the Illuminating Engineering Society *Lighting Guide*. Lighting systems must be kept in good working order.

## CLEANLINESS

Poor housekeeping standards increase the risk of injury, disease, fire and damage and reduce efficiency. So it makes economic sense — and improves safety — to maintain high standards of housekeeping and cleanliness. Supervisors should carry out regular housekeeping inspections and ensure that high standards are maintained.

## CHEMICALS

It is the responsibility of management to ensure that adequate information is obtained on every substance used in facilities under its control. This enables an assessment of the hazards associated with every substance to be made and the necessary precautions identified as required by the Control of Substances Hazardous to Health Regulations 1988 (COSHH), amended in 1991 and 1992. Chemical safety data sheets should be available for all hazardous materials handled on a facility.

Materials should only be accepted on to a site if they are properly and adequately packaged and labelled in accordance with the Chemicals (Hazard Information and Packaging) Regulations 1993 and good industrial practice, bearing in mind the hazards of the material concerned. A material safety data sheet (MSDS) should be held in every area that handles the material, and should be available for consultation by all employees.

Appropriately qualified personnel should carry out an assessment of the activities associated with the materials used on site in accordance with the Control of Substances Hazardous to Health Regulations 1988. They should also assess personal protection required in accordance with the Personal Protective Equipment at Work Regulations 1992. To assess the hazards of substances it is necessary to identify the toxicity of all the substances used on site, the exposure level, and the number of personnel and the length of time they are exposed to the substances.

The assessment of personal protection is achieved by reviewing each task carried out using a Hazards Identification Record Sheet as shown in Figure 7.1 on pages 74–75, ticking the box for each part of the body which could be affected by the relevant hazard. The initial list of hazards consists of general items which are covered by the PPE Regulations, whilst the list marked 'Specific Legislation' covers requirements of one or more of the following;

- Control of Lead at Work Regulations 1980;
- Ionizing Radiation Regulations 1985;
- Control of Asbestos at Work Regulations 1987;
- Control of Substances Hazardous to Health Regulations 1988;
- Noise at Work Regulations 1989.

Employees who handle hazardous substances also have a responsibility to use the necessary precautions and protection.

Having identified the hazards involved in the task, record the controls that are in place to prevent the hazards causing injury (in a table as shown in Figure 7.2 on page 76) and assess their effectiveness as high, medium or low. Then assess the exposure by looking at the number of people exposed and the likely frequency of failure, again using high, medium and low assessments. This also applies to the likely consequences of failure. The risk level is a combination of the exposure and consequences. A full discussion of this method of assessing risk is given in Chapter 5. If the risk level is high, then at least two independent and highly effective controls must be in place. If the risk level is medium, there must be at least one highly effective and one medium effective control.

Personal protective equipment (PPE) should be regarded as no more than a medium effective control. If the standard of controls described is not available then additional controls should be introduced. Finally, if appropriate, additional PPE should be recommended.

Having identified the current and required personal protective equipment, review the suitability of the current provisions with regard to PPE. First, list the various PPE required for each task and identify all current models supplied on a table as shown in Figure 7.3 on pages 76–77. Then assess its suitability to provide the protection required and compatibility with other PPE which must be used simultaneously. Finally, assess the user acceptability, the specification, storage and inspection/maintenance requirements.

**RESPIRATORY PROTECTION**

It is necessary to take all reasonably practical steps to minimize the risk of respiratory health hazards. This is achieved using, in order of preference, substance substitution, engineering controls, ventilation and finally respiratory protection.

The management of a facility should establish an effective respiratory protection programme. This programme should define the various respiratory hazards for each work place and activity and the acceptable protective equipment.

Wherever possible, representatives of the work force should participate in the selection of the equipment to be used.

Where an atmosphere could be dangerous to life, the minimum acceptable standard should be either a self-contained breathing apparatus in the pressure demand mode or an airline pressure demand respirator with an auxiliary emergency supply bottle of sufficient capacity to permit escape from the area.

Breathing air should not be taken from a plant air system which could

| | Head | | | | |
|---|---|---|---|---|---|
| | Head | Ears | Eyes | Respiration | Face |
| Body | | | | . | |
| Falling objects | | | | | |
| Blows, cuts, crushing | | | | | |
| Vibration | | | | | |
| Slips, trips, falls | | | | | |
| Scalds, heat, flame | | | | | |
| Cold | | | | | |
| Cold burns | | | | | |
| Immersion | | | | | |
| Non-ionizing radiation | | | | | |
| Electrical | | | | | |
| **Specific legislation:** | | | | | |
| Noise | | | | | |
| Ionizing radiation | | | | | |
| Dust | | | | | |
| Fume | | | | | |
| Splashes, spurts | | | | | |
| Gases, vapours | | | | | |

Figure 7.1    Hazard identification record sheet.

be connected to the process system. It can be taken from an instrument air system, provided it passes through an absorbent bed to remove organic vapours and carbon monoxide. The breathing air supply should be fitted with a high temperature alarm and the coupling should be incompatible with other hose couplings. Breathing air should meet the requirements of BS 4275: Recommendations for the Selection, Use and Maintenance of Respiratory Protective Equipment.

Personnel who use respiratory protection should be medically fit to do

| Upper limbs | | Lower limbs | | Various | | |
|---|---|---|---|---|---|---|
| Hand | Arm | Foot | Leg | Skin | Trunk | Whole |
| | | | | | | |
| | | | | | | |
| | | | | | | |
| | | | | | | |
| | | | | | | |
| | | | | | | |
| | | | | | | |
| | | | | | | |
| | | | | | | |
| | | | | | | |
| | | | | | | |
| | | | | | | |
| | | | | | | |
| | | | | | | |
| | | | | | | |
| | | | | | | |
| | | | | | | |

| Controls (including PPE) in use | Effectiveness |
|---|---|
|  |  |
|  |  |
|  |  |
|  |  |
| No. of employees affected | High/medium/low |
| Frequency of failure | High/medium/low |
| Consequences of failure | High/medium/low |
| Assessment of risk | High/medium/low |

Figure 7.2    Assessment of controls record sheet.

| Tasks | PPE required | British Standard | Current models supplied | Suitability | |
|---|---|---|---|---|---|
|  |  |  |  |  |  |
|  |  |  |  |  |  |
|  |  |  |  |  |  |
|  |  |  |  |  |  |
|  |  |  |  |  |  |
|  |  |  |  |  |  |
|  |  |  |  |  |  |
|  |  |  |  |  |  |
|  |  |  |  |  |  |
|  |  |  |  |  |  |
|  |  |  |  |  |  |
|  |  |  |  |  |  |
|  |  |  |  |  |  |

Figure 7.3    Review of PPE provisions.

so and be trained in its use and how to fit it correctly. Respiratory protection should be regularly inspected, tested, maintained and cleaned.

## RADIATION
There are two types of radiation:
- non-ionizing radiation — such as infrared and ultraviolet;
- ionizing radiation — such as gamma and x-rays.

### NON-IONIZING RADIATION
Sources of non-ionizing radiation should be clearly identified since they can induce sparks. Very short wave radiation can cause heating of body tissue and access to such equipment must be strictly controlled.

### IONIZING RADIATION
The management of the use of ionizing radiation must be strictly controlled in accordance with the Ionizing Radiation Regulations 1985. These require the

| Compatibility | User acceptability | Storage requirements | Inspection/ maintenance requirement | Recommendation |
| --- | --- | --- | --- | --- |
| | | | | |
| | | | | |
| | | | | |
| | | | | |
| | | | | |
| | | | | |
| | | | | |
| | | | | |
| | | | | |
| | | | | |
| | | | | |
| | | | | |
| | | | | |

development of a Radiation Protection Policy and local rules governing the activities involving radioactive substances. This involves the appointment of a suitably qualified Radiation Protection Advisor (RPA) to provide advice and guidance and any necessary Radiation Protection Supervisors (RPSs) to ensure compliance with the policy and local rules. A radioactive materials audit should be carried out annually by the RPA accompanied by the relevant RPSs, and a contingency plan developed to plan how an incident involving ionizing radiation will be dealt with without risks to the health and safety of personnel. Contractors working on company premises should work under the company's local rules unless the RPA has agreed any deviation. All transport of radioactive materials must conform to the relevant international agreements on the transport of dangerous materials by air, sea, road and rail with regard to quantities, packaging, labelling and notification. Radioactive materials should only be stored in a properly designated store which is adequately segregated from hazardous areas, soundly constructed, clearly labelled and kept locked at all times.

## NOISE AND VIBRATION

All practical steps should be taken to keep noise levels as low as possible and minimize the exposure of personnel to hazardous noise levels by providing 'quiet' equipment, sound proofing or (as a last resort) hearing protective equipment. Noise surveys should be carried out regularly and noise contour maps produced. Where noise levels are above the acceptable level of 85 dBA for 8 hours, octave band analyses should be carried out and appropriate noise reduction measures instituted. Where noise levels are above 80 dBA, suitable hearing protection should be provided and employees trained in the hazards of noise and the correct use of the protection provided.

Management should also have a vibration survey carried out to identify any areas where hazardous levels of vibration exist. Access to these areas must be properly controlled. Any hand-operated tools which create dangerous levels of vibration should be repaired or replaced forthwith.

## MANUAL HANDLING

A manual handling programme should be developed in line with the Manual Handling Operations Regulations 1992 to reduce the risk of injury due to handling or lifting operations. A survey should be carried out to identify all manual handling and lifting activities. Assess the level of risk by carrying out reviews in a logical progression, starting with a preliminary review to identify potentially hazardous activities and becoming progressively more detailed as the higher risk jobs are identified. Figure 7.4 illustrates this.

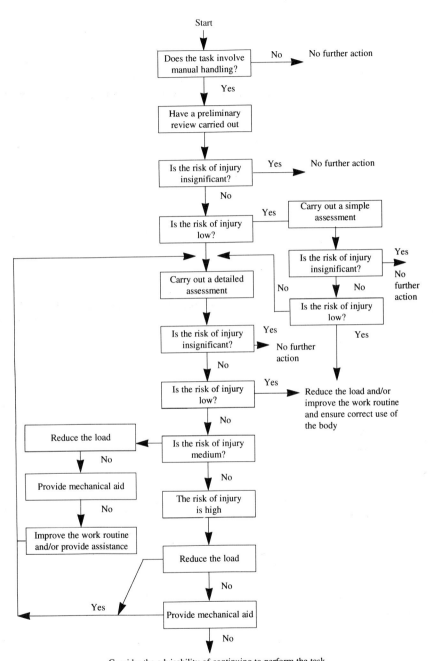

Figure 7.4    Manual handling risk assessment flow diagram.

| Date: | | Location: | |
|---|---|---|---|
| Review team: | | | |
| Are manual handling tasks carried out? | | Yes/No | |
| **Assessment/tasks:** | | | |
| **Task** | **Describe the load** | **Movements involved** | **Risk of injury*** |
| | | | |
| | | | |
| | | | |
| | | | |
| | | | |
| | | | |
| | | | |
| * Rate risk of injury as insignificant, low, medium or high | | | |

Figure 7.5    Preliminary review of manual handling tasks.

PRELIMINARY ASSESSMENT

Divide the facility into suitable discrete work areas and identify all the tasks in the area which involve manual handling, as shown in Figure 7.5. For each load describe the load involved, its weight and shape, balance and so on, along with the movement involved in handling it. From this, rate the risk of injury as insignificant, low, medium or high. If the risk of injury is insignificant, no further action need be taken. If the risk of injury is low then a simple assessment should be carried out, whereas if the risk is medium or high, a detailed assessment is required.

SIMPLE ASSESSMENT

Using the table shown in Figure 7.6, record the task (for example, type of movement, speed and repetition), the load (for example, weight, volume and awkwardness) and the work environment (for example, space, floor condition, lighting and temperature). Finally describe any special capabilities needed by the operator (for example, length of reach, fitness or training). Then consider how the risk of injury could be reduced.

DETAILED ASSESSMENT

A detailed assessment should be carried out for all tasks where the preliminary assessment indicates a high or medium level of risk of injury. The assessment should be assisted by a manual handling expert. If the risk is confirmed as medium or high then it may be necessary to obtain the assistance of a mechanical handling equipment expert. The detailed assessment should be carried out for each task individually using Figure 7.7 — see pages 82–83 — with each factor

| Date: | | Location: |
|---|---|---|
| Review team: | | |
| Task: | | |
| Describe the task: | | |
| Describe the load: | | |
| Describe the environment: | | |
| Describe any special capabilities needed by the operator: | | |
| | **Yes/No** | **If Yes, describe** |
| Can the load be reduced? | | |
| Can mechanical assistance be provided? | | |
| Can human assistance be provided? | | |
| Can the work routine be improved? | | |
| How can the correct use of the body be ensured? Describe | | |
| What priority is required? — high/medium/low | | |

Figure 7.6    Simple manual handling risk assessment for tasks with low levels of injury risk.

81

| Date: | | Location: | | | |
|---|---|---|---|---|---|
| Review team: | | | | | |
| | | **Risk level** | | | |
| | Yes | Low | Medium | High | Comment |
| **Tasks — do they involve:** | | | | | |
| • holding loads away from trunk? | | | | | |
| • twisting? | | | | | |
| • stooping? | | | | | |
| • reaching upwards? | | | | | |
| • large vertical movement? | | | | | |
| • long carrying distances? | | | | | |
| • strenuous pushing or pulling? | | | | | |
| • unpredictable movement of loads? | | | | | |
| • repetitive handling? | | | | | |
| • insufficient rest or recovery? | | | | | |
| • a work rate imposed by a process? | | | | | |
| • variations in levels? | | | | | |
| • hot/cold/humid conditions? | | | | | |
| • strong air movements? | | | | | |
| • poor lighting conditions? | | | | | |

Figure 7.7   Detailed manual handling risk assessment for tasks with medium levels of injury risk.

| | Yes | Risk level | | | Comment |
| --- | --- | --- | --- | --- | --- |
| | | Low | Medium | High | |
| **Loads — are they:** | | | | | |
| • heavy? | | | | | |
| • bulky/unwieldy? | | | | | |
| • difficult to grasp? | | | | | |
| • unstable/unpredictable? | | | | | |
| • intrinsically harmful? (eg, sharp, hot) | | | | | |
| **Working environment — are there:** | | | | | |
| • constraints on posture? | | | | | |
| • poor floors? | | | | | |
| **Individual capability — does the job:** | | | | | |
| • require unusual capability? | | | | | |
| • hazard those with a health problem? | | | | | |
| • hazard pregnant women? | | | | | |
| • call for special information/training | | | | | |
| **Other factors:** | | | | | |
| **Overall level of risk of injury: Insignificant/low/medium/high** | | | | | |

assessed individually for its level of risk of injury so that the overall level can be assessed. Once the overall assessment is complete, if the level is insignificant no further action is needed; if it is low or medium then ways of reducing it should be reviewed. If the level is high then the risk level must be reduced, or consideration given to whether it is wise to continue to carry out the task. Once ways of reducing the risk, in the case of low or medium levels, have been identified, the overall risk level should be re-assessed. The level must be reduced to at least the low category.

As part of a manual handling programme employees should be instructed and trained in the correct technique to be used when handling loads.

### EYE PROTECTION

It is essential that no-one suffers eye injury due to work activities. To achieve this everyone should consider the potential for eye injury before starting any job and throughout its course. If an eye hazard is apparent, take steps to eliminate it or protect the exposed personnel by the use of eye protection. Eye protection should be specified in line with the national or international standard.

### DISPLAY SCREEN EQUIPMENT

Display screen equipment (DSE) should meet the requirements of the Health and Safety (Display Screen Equipment) Regulations 1992. They involve:

- identifying all DSE users;
- offering all users free eyesight tests as often as necessary;
- providing any users who require special spectacles to allow them to operate DSE with free spectacles;
- ensuring that a self assessment of all users' work stations is carried out annually and that a detailed assessment is carried out on the work station of any user suffering difficulty;
- training all users to use the DSE safely.

All new DSE work stations must meet the requirements of the schedule to the regulations.

# 8. HAZARD AWARENESS

## ACCIDENT VERSUS HAZARD

The possible results of an accident are injury, property damage or a near miss (Figure 8.1).

Think of a spanner falling off a roof. It might hit someone below — an injury. It might hit a car — property damage. It might miss everything and bounce off the ground — a near miss. It is the same accident but the results are very different. This is why an accident is defined as an unplanned event which may or may not result in injury or property damage. Many studies have been carried out into the ratios between the various outcomes of accidents. One of these has produced the accident triangle shown in Figure 8.2.

Thus, although there are very few injuries, many significant property damage and near miss accidents occur and a vast number of hazards.

What is the difference between an accident and a hazard? A hazard is a situation in which danger exists and it is possible for an accident or incident to occur. Figure 8.3 shows the effect of hazards.

## DOMINO MODEL

Hazards can be classified as unsafe acts or unsafe conditions (Figure 8.4). But hazards — that is, unsafe acts and/or conditions — can exist for long periods of

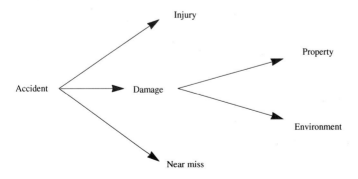

Figure 8.1    The possible results of an accident.

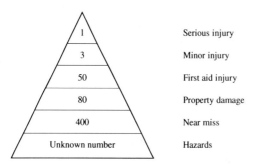

Figure 8.2    Accident triangle.

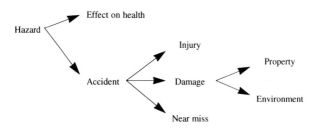

Figure 8.3    The effect of hazards.

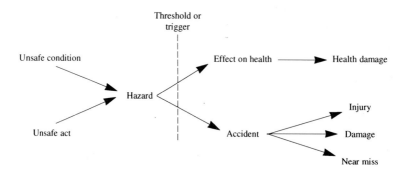

Figure 8.4    Unsafe acts and conditions.

time without an accident occurring. It appears that some kind of trigger or crossing of a boundary or threshold allows the accident to happen. A hole in a floor can exist for days with everyone stepping around it but one day someone rushing to get a job done does not notice it, steps in it and twists an ankle.

## HAZARD CAUSES

Hazards are caused by unsafe acts and omissions or by unsafe conditions. Unsafe conditions are inadequacies in either the situation or the systems of work, whilst unsafe acts are personal failings which are a risk to health and safety.

Hazard identification is therefore a fact-finding process to identify the components of a hazard under these two headings.

Unsafe acts of commission are acts that deviate from a specified or generally accepted safe way of performing a task. Such acts increase the likelihood of an accident, occupational ill health or disease. For example:

- unauthorized use of machinery and equipment;
- operating machinery and equipment at unsafe speeds or overloading;
- nullifying safety devices;
- using defective equipment;
- unsafe use of equipment, home-made lash-ups;
- taking up unsafe positions;
- attempting repair or maintenance of energized or moving equipment;
- riding on hazardous equipment;
- horseplay.

Unsafe acts of omission are failures to conform to the requirements of the activity in a manner that increases the likelihood of an accident, occupational ill health or disease. For example:

- failure to follow prescribed procedures;
- failure to conform to legislative requirements;
- failure to make secure;
- failure to warn or signal;
- failure to use personal protective equipment;
- failure to instruct and train.

Unsafe acts and omissions can be consolidated into:

- inappropriate attitudes;
- inadequate knowledge; } Consequence of inadequate arrangements
- inadequate skill;
- inadequate supervision;

87

- failure to do 'something':
— follow procedures;
— follow work method;
— non-use of protective devices, etc.

Unsafe conditions can be manifested in any aspect of the working environment. For example:

- environmental conditions (light level, noise and vibration, dust, fumes);
- inadequate access to and egress from fire exits, parts of the process, etc;
- product inherently hazardous (hot, cold, flammable);
- materials/substances used in the manufacturing process are hazardous;
- inadequacies in the layout of plant and equipment;
- inadequacies in the design of plant and equipment (including guarding);
- condition of plant and equipment (including guarding);
- hazardous work procedures;
- hazardous work methods;
- inadequate arrangements and methods used for giving information, instruction, training and supervision;
- poor housekeeping.

The relationship between an accident or adverse effect on health and unsafe acts and conditions is shown in Figure 8.5. However it is a fact that hazards can exist, often for extremely long periods of time, without an accident or apparent effect on health. So it seems as if some sort of hypothetical threshold has to be crossed. The factors which cause that threshold to be crossed are complex and not completely understood. Since an accident or effect on health can only occur if a hazard exists, then a hazard must — in a sense — provide the impetus to breach the threshold and cause an accident or adverse effect on health. An analogy is an elastic membrane — for example, a rubber balloon. The membrane can be stretched by an applied pressure but it returns to its original state if and when the pressure is removed. If the pressure becomes so great that the elastic limit of the membrane is exceeded, then the membrane ruptures.

A hazard provides a continual threat or risk of an accident/effect on health, and the only practical way of preventing accidents and effects on health is to remove or at least effectively control the hazard.

## HUMAN FACTORS MODEL

As the name suggests, the human factors model (see Figure 8.6 on page 90), is based on the identification of the reasons why the personnel behaved in the way that allowed the hazard to occur. The model is based on the fact that human

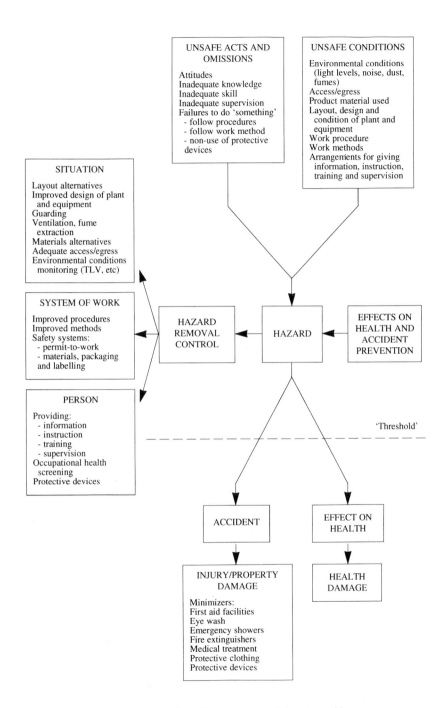

Figure 8.5    The relationship of unsafe acts and conditions to accidents.

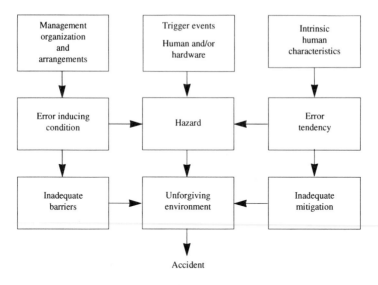

Figure 8.6    Human factors accident model.

behaviour is influenced by two factors — any intrinsic error tendencies present in the individual and any error-inducing conditions applied to the person. When either or both of these are present and a trigger occurs, then a hazard will exist. If the environmental conditions are unforgiving whilst the hazard exists, then an accident will occur.

The intrinsic error tendencies of a person are inherent characteristics of that individual due to physical and physiological make-up, limitations, etc.

The error-inducing conditions are external to the person but influence performance by creating mismatches between the person and the requirements needed to do the job safely — for example, training, skills development, motivation, stress, supervision, communications, peer pressure, company culture. The unforgiving environment is the inadequacy of the facilities, barriers, controls and procedures which allow the accident to develop and the ineffectiveness of the mitigating factors, such as protection facilities.

Another way of looking at the tendency for people to be involved in an accident is depicted in Figure 8.7. This model shows the various factors which influence what a person does in any situation. If enough of the factors are against safety then a hazard will develop, either by the person allowing an unsafe situation to appear or by carrying out an unsafe act.

# HAZARD RECOGNITION

## INSPECTIONS

On every facility there are hundreds, if not thousands, of items like pipes, hoses, wires, cables, chains, pulleys and belts that must wear out at some point. Normal wear and tear may bring about gradual deterioration that may be detected before any personal harm, property damage or work interruption occurs. On the other hand, failure of many of these items can take place suddenly and involve circumstances that present harmful exposure to people and property.

In addition to the undesired exposures created by many items as they wear out, there is also the ever-present loss potential from items that have been damaged or rendered inefficient by abuse and misuse. The disorderly arrangement of materials and equipment through other poor work habits adds another avenue for potential loss.

## UNPLANNED INSPECTIONS

There are two basic types of inspection — the unplanned inspection and the planned inspection. The first type comes so naturally that it needs very little explanation as every supervisor carries it out automatically as he goes about his normal activities. For most efficient follow-up, many people have learned to jot

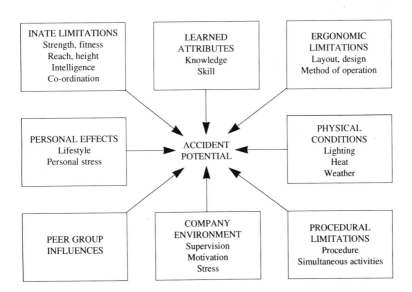

Figure 8.7    Human error influences.

91

down substandard conditions as they 'spot' them, in a pocket-sized book. These notes later serve as memory-joggers and are crossed off as they are actioned.

Unfortunately informal inspections tend to pick up only the very obvious problems and those that occur on or in the vicinity of the route. This is not meant to indicate that the unplanned or informal type of inspection does not make an important contribution. It is necessary, however, to emphasize clearly that the informal method should supplement the planned or formal type of inspection.

## PLANNED INSPECTIONS

The general inspection involves a walk-through of an entire area, with the inspector looking for anything and everything that could potentially degrade the operations. It is not uncommon for inspections to be made by small teams, preferably including a representative from the work force. The general inspection is most frequently made monthly or bi-monthly, with all items recorded accurately and classified by degree of risk. The frequency of inspections should give reasonable time for personnel involved to accomplish the remedial work on items reported, yet be frequent enough to keep control of the hazards.

### How to make an inspection

(1)  Look for off-the-floor and out-of-the-way items. Get the 'big picture' of the whole area. It is usually items that are off the beaten track that cause problems. Remember, informal inspections pick up many of the obvious hazards. Spend a good amount of time looking for things that will be missed in everyday operations.

(2)  Systematically cover the area. Allot a certain amount of time to do the job. In order to cover everything in a methodical and thorough manner, it might be worth walking through the area for a 'once-over-lightly' view, to decide on the best path for a proper and complete inspection. Often an inspector will misjudge the amount of ground he has to cover.

(3)  Describe and locate each item clearly. Much time can be wasted answering questions and revisiting areas after an inspection report is submitted because descriptions of items or their locations were not clear and accurate. Use established names and wall or column markings, if available, to pinpoint locations. When in doubt, ask someone — don't rely on memory.

(4)  Follow up urgent items quickly. When anything is discovered that represents a serious risk or potential danger, take action immediately. There are usually many temporary immediate measures that can be taken to reduce the risk until more suitable permanent correction can be instituted — for example, placing an improvized guard rail around an open pit, or blocking off an area around a pile of materials.

## HAZARD REPORTING

Hazard reporting is a formal method for anyone who identifies a near miss or a hazard to report it. It usually takes the form of a short written report that the person uses to identify the hazard, its location and what action, if any, has been taken to remove or contain it. The form is given to a supervisor who ensures that it is given to the person responsible for the area. It is useful if the form has space on it for the 'responsible supervisor' to assess the risk potential and record what action has been taken or will be taken to remedy the hazard. A copy of the completed form is returned to the originator and any outstanding remedial actions entered into a tracking system to monitor implementation. This system can be used to trigger a useful recognition programme.

## CRITICAL TASK ANALYSIS

The basis of critical task analysis is that certain jobs have the potential to cause serious safety hazards if they are not done correctly, whilst other jobs cannot have a major impact on safety no matter how they are done. To determine which jobs are 'critical' consider:

- past injuries and incidents;
- new or unusual jobs;
- potential for injury or damage.

Having identified the critical jobs, break them down into steps and then consider people, equipment, materials and environment and develop recommendations where appropriate to eliminate or contain the hazards.

### PEOPLE

- What contacts with energy, physical or chemical, are present that could result in injury or illness?
- Are there contacts that could cause fire or explosion?
- Does the worker understand and follow all rules, regulations and precautions?
- Has proper personal protective equipment been provided? Is it being utilized properly?
- Are the right number of people doing the job?
- Is there any idle time for workers that could be used more gainfully?

### EQUIPMENT

- Are the tools and equipment being used best suited to this job from all aspects (safety, quality, production)?
- Can tools be provided that would improve efficiency and safety?

- Can mechanical or power tools be applied more economically than hand tools?
- Is machinery and equipment being used to its maximum safe capacity?
- Is all equipment in proper operating condition?
- Is there a less costly piece of equipment that would do the same job properly?
- Are all tools readily available and properly positioned for most effective work?

MATERIAL
- Can better, safer, less expensive or less scarce material be substituted?
- Can waste be cut down in any way?
- Are there any uses for scrap or waste?
- Is material being transported and handled in the most efficient manner?
- Is the right amount of material at the job site?
- Is there another product that could do the same job at less cost?

ENVIRONMENT
- Are working areas and related storage areas clean and orderly?
- Is junk occupying space that could be used to more advantage by people, equipment or material?
- Is there anything in the environment that you would consider unnecessary to the task at hand?
- What in the environment could be changed or altered to improve conditions, atmosphere or general work climate for people/equipment/material?

**INCIDENT RECALL**

Incident recall is a technique in which a group of employees sit down under the leadership of a supervisor and brainstorm occurrences which have actually happened. Obviously it is best to focus on a specific piece of equipment or activity that the group is familiar with — and it is essential that absolutely no blame is attached to anyone who raises a problem or hazard, no matter how it was caused. Remember these are hazards and occurrences that no-one knew about before, and hence no remedial action was taken and the event could occur. It is much more important to stop the recurrence than to take action against anyone who was at fault. Unless this point is accepted and followed, incident recall will not be effective; the points raised must be carefully sieved to remove any in which individual fault is apparent.

Prior to holding the session the supervisor should think through the activity or use of equipment and note down potential sources of hazard to stimulate the thinking of the group.

As the events and hazards are brought up, initially merely record them. Do not comment on their potential. Once the discussion starts to dry up, put the ideas into some sort of systematic order and consider their risk level using the technique already described. Then take the high risk items and agree with the group what actions will be taken immediately to ensure safety. After the session, make out a report indicating what short-term action has been taken and all recommendations for long-term action. The report should be passed up the management line for approval and action. Equally, senior management should report back down the line when the long-term actions are being taken and the supervisor should feed this back to the group and track implementation.

Learning Resources
Centre

# 9.  CONTRACTOR OPERATIONS

Many manufacturing companies consider that the contracting companies they use have poorer safety performances than themselves. The costs of this poor performance are passed on to the user in the form of higher bids and longer schedules. In addition, psychological and cultural effects of accidents affect the company as much as the contractor. The company therefore benefits from a successful contractor safety programme in many ways. Company employees benefit from a safer work environment as do the contractor's employees. With improved contractor safety performance, the company benefits from lower costs through more competitive contractor proposals and fewer civil liability suits which can potentially involve the company. Fewer accidents allow better continuity in the contractor's work, resulting in a better-quality product for the company. The overall quality of the contractor community is enhanced as safety attitudes and performance improve. The company realizes an improved public image as it develops a safer work environment for all persons at its facilities.

The benefits of a successful programme to the contractor's organization extend from the individual employee on site to top management. The safety programme maintains the health and wellbeing of the individual employee and reduces the potential for personal injury, loss of income resulting from temporary or permanent disabilities, and subsequent family disruption. Clearly the site employee is a prime beneficiary of a good safety programme. For the contractor, a successful safety programme results in fewer accidents which translate into improved productivity and ultimately lower operating costs. With lower operating costs and a good safety record, the contractor can potentially be more competitive in the market place.

The means through which improved contractor safety can be achieved are:
- careful selection of contractors;
- a contract precisely defining the contractor's responsibilities;
- adequate training of the company's own and contractor personnel;
- proper supervision and safety auditing;
- an efficient reporting system;
- performance evaluation;
- a good example set by company personnel.

## RESPONSIBILITIES

In matters of safety each party involved in operations bears its own individual responsibilities. Each contractor should be aware that it may be held responsible if an accident happens as a result of its employees' activities. In order to prevent accidents, however, the company (as the contractor's principal) also has a responsibility of its own: the company should require that the contractor abide by specific safety standards and rules which have been defined in the contract and wherein the contractor's responsibilities in regard to safety should be clearly set out.

The preferred allocation of responsibilities under the contract can be summarized as follows:

- the contractor has a primary responsibility for the safety of its own operations, as defined in the contract between the company and the contractor;
- the company is responsible for clearly instructing the contractor on its safety requirements as well as for enabling the contractor to abide by these safety requirements, and for supervising the contractor's performance in respect of safety;
- within the company's organization, the primary responsibility for all aspects of safety, including the contractor's safety, should lie with line management.

Because the basic responsibility for contractor safety lies with the contractor, contractor employees must be properly trained, instructed, equipped and supervised as required by the contract agreement. The company may, in certain circumstances, assist the contractor in the fulfilment of specialized training obligations. But this does not relieve the contractor of the responsibility to ensure adequate employee safety training.

The contractor is also responsible for ensuring that subcontractors have proper safety training and that their safety programmes satisfy the contractual requirements between the company and the contractor.

The company should designate a site representative who is responsible, amongst other things, for monitoring the contractor safety performance. Even though the contractor has the basic responsibility for its employees' safety, the company representative also has the responsibility to prevent accidents and injuries by monitoring and analysing the safety of the contractor's activities, and setting an example by observing and practising safe work habits.

Under no circumstances should an employee or contractor be assigned a task without sufficient training to accomplish it safely and efficiently.

The success of the contractor safety programme depends heavily on the commitment and motivation of the company representative. Meetings between the company representative and the contractor should be held regularly to make sure everyone on the job site understands that safety is of paramount concern and an important factor in the success of the entire project.

97

It is imperative to state clearly in the contract documents the acceptable standards pertaining to safety performance and the responsibilities allocated to each party. The following list gives examples of the responsibilities which might need to be clarified in the documentation. Not all of them apply in every case and the list is not exhaustive. It is only a guide:

- the provision of materials and equipment (including the necessary safety equipment and associated approval certificates where appropriate) and arrangements for instruction in their use;
- the provision and use of rescue/emergency equipment and first aid equipment, access to medical facilities and arrangements for emergency medical assistance;
- the distribution among the contractor's employees of any information concerning health and safety hazards, safety practices or procedures, whether verbal or written, when conveyed to the contractor by the company. This includes, for example, the issue of work permits;
- the arrangements for health surveillance and monitoring of the working environment;
- the maintenance of statutory records or registers relating to the work in hand or to individual employees;
- the issue of any permit, written instruction, identification, certificate or other document either by statute or by the company. Individual permit-to-work systems vary according to circumstances and the degree of risk. It is vital that the workings of any permit system are fully understood by all interested parties and that persons are appointed to oversee full adherence to the system;
- the performance and evaluation of any atmospheric or any other test procedure required either by statute or by the company;
- the specification and maintenance of means of access to, and exit from, a contractor's place of work;
- arrangements for the collection, treatment and disposal of waste in accordance with current legislation;
- the recording and reporting of accidents, injuries, incidents, near misses and instances of occupational ill health and their subsequent investigation;
- arrangements for work force consultation and involvement;
- the Substance Abuse Policy to be applied on the site;
- the provision of safety personnel;
- the safety training required.

## ENFORCEMENT

The company should not tolerate poor safety practices performance by contractors working for the company. Contractors must be made to understand that the company is serious about safety. Failure to comply with the established safety rules and procedures should result in an immediate cessation of work until the violation is corrected, termination of the contract, and/or removal from the company approved bidders list.

## CONTRACTOR SAFETY EVALUATION

It is essential to select contractors who have the ability and willingness to perform the work. The selection criteria should include an assessment of the contractor's record in appropriate health and safety matters and the contractor's ability to carry out the work safely.

A full competitive tender exercise can be divided into two stages:
- the preparation of the bid list for approval (pre-qualification);
- the evaluation and selection of the preferred contractor, including bid clarification sessions with contractors.

By applying a careful safety screening process at the bid list preparation stage — that is, eliminating those contractors with a poor past safety performance record, or lack of commitment towards the enhancement of safety in its operations — the problem of being forced at the evaluation and selection stage to accept the lowest bidder with an unacceptable safety record is largely overcome. It should be the aim of the company to be fully satisfied that the contractors which are eventually invited to bid can perform the work safely. If there is insufficient information to carry out an assessment at the bid list preparation stage, it is acceptable to include the contractor on the bid list. This must be clearly flagged, however, so that thorough checks can be carried out at the bid assessment stage.

A guide to the programme elements to be used when selecting contractors for various types of contracts is given in Figure 9.1 on page 100.

### TENDER QUESTIONNAIRE

All requests for tenders which involve the provision of labour should include a safety questionnaire to be filled in by the tenderer and submitted with the tender. The questionnaire covers all aspects of the potential contractor's health and safety competence.

### *Assessment of management attitude*

Evidence of good managerial competence on the part of the contractor should be sought. The professional calibre of the senior directors or management staff often reflects the overall performance of the contracting company, but it is also important to establish the competence of the individuals who will manage the

99

Does contract involve?

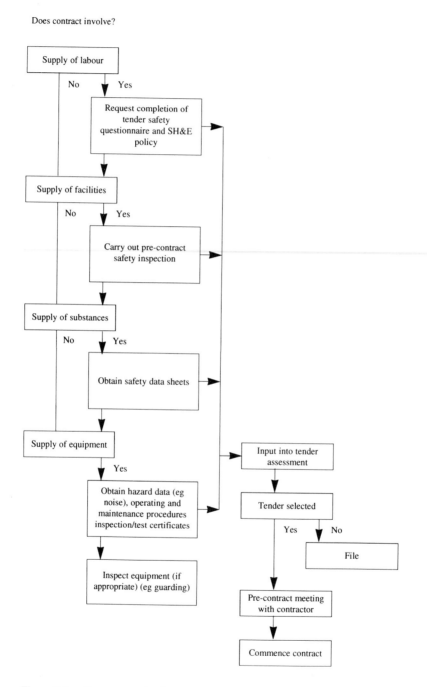

Figure 9.1    Contractor selection programme.

contract. The contractor should be able to describe his management or supervisory structure and indicate how this fits into the overall plan for ensuring the health and safety at work of affected personnel. The contractor should also be able to demonstrate that the individuals responsible for overseeing health and safety matters are competent.

Similar consideration should be given by the contractor to the appointment of subcontractors. In this respect, appointment of subcontractors may follow discussion, agreement and exchange of relevant information between the contractor and company. Although a contractor will often reserve the right (by virtue of the contract) to approve the employment of any particular subcontractor, it should be possible to demonstrate that selection procedures used are not prejudicial to the overall efficient management of the work nor the interests of health and safety.

The involvement of the employee work force in the safety programme should be checked. Contractors should be able to demonstrate this through policies and involvement — for example, safety representation and committees and anti-intimidation policies and safety motivation programmes.

### Assessment of health, safety and environmental policy

The contractor is normally required by statute to have a suitable safety policy, and both contractor and company should ensure that their respective safety policies are compatible.

The details of a safety policy depend upon the individual company organization and systems. Contractors may need to review and revise their safety policies more frequently than operators at permanent locations, since each new work location or activity presents its own particular problems. It may be convenient for a contractor to have a general safety policy statement, applicable throughout the company, with a separate specific statement for each work location. Thus the complete safety policy would consist of a general statement and one or more specific statements of policy.

The health and safety policies should be backed up by suitable programmes, procedures and practices. Evidence should be obtained on their coverage and depth.

### Assessment of health and safety performance

The company should review with the contractor management the contractor's health and safety record. In some rare cases, contractors may be legally required to hold licences, or similar, for the type of work they carry out, and this matter too should be fully pursued.

The history of accidents, instances of occupational ill health or dangerous incidents in which the contractor has been involved are also an important consideration. The contractor should be prepared to provide information on such incidents.

The contractor's safety and health programmes should be reviewed to confirm their adequacy.

*Employee competence*

The contractor should be able to demonstrate that its employees are competent by suitable selection and training programmes, especially in critical skills.

*Emergency preparedness*

The emergency plans, facilities and procedures of the contractor should be reviewed to ensure that they are adequate to deal with foreseeable scenarios.

Returned questionnaires should be assessed using a suitable assessment system and the results input into the company tender assessment procedure.

**PRE-CONTRACT INSPECTION**

A contract for the provision of a facility is not normally signed until after a physical inspection of the facility has been carried out. The inspection should be carried out against a suitable check list to establish any sub-standard aspects of the facility. The contract for the selected facility should include a requirement to correct essential sub-standard aspects before the start of the contract or by an agreed time as appropriate. Other non-critical sub-standard aspects should be drawn to the attention of the contractor with a recommendation that they be attended to.

**SAFETY DATA SHEETS**

Every request for a tender for the provision of a substance or material should include a requirement for the inclusion of a safety data sheet with the tender. The safety data sheet should include information on:
- the health and environmental hazards of the material;
- the precautions to be taken during transport and storage;
- the precautions to be enforced when using it;
- the hazards involved in emergency situations;
- the actions to be taken in event of spillage.

The recent European Commission Directive on Chemical Hazard Information and Packaging provides an excellent standard safety data sheet.

The overall hazard of all technically suitable materials or substances should be assessed and input into the tender assessment to assist in the award of the contract.

## HAZARD DATA SHEETS

Every request for tender for the provision of plant or equipment should include a requirement for the submission of a hazard data sheet. This includes information on:

- physical hazards — for example, pinch points;
- health hazards — for example, noise and radiation;
- guards and fitted protection;
- any protection to be provided by the company.

The hazards and requirements for safe operation should be assessed and input into the tender assessment.

Where appropriate, a physical inspection of the proposed equipment should be carried out to ensure that all guards and protection provided by the contractor are adequate and up to standard. The contract should include a clause requiring any deficiencies to be corrected before the goods are despatched to the company.

## TENDER ASSESSMENT

Wherever possible the safety, health and environmental standards of all tenderers should be assessed using the techniques already described and used as one of the measures to assess the various tenders when selecting the contractor. Any tender which does not meet the company's standard should not be awarded a contract without the express agreement of a designated senior manager and a clear definition of how the safe execution of the contract will be achieved.

## PRE-AWARD MEETING

Where appropriate, a pre-award meeting should be held with the short-listed contractors to clarify any uncertainty about their safety and health performance, competency or commitment. In addition, clarification can be obtained about any possible weaknesses and a commitment obtained about their correction in event of the award of the contract. Finally, agreement can be obtained in broad terms about the respective responsibilities of the company and the contractor and how the detailed arrangement will be finalized. Specific items to be covered include, but should not be limited to:

- provision of safety and emergency equipment and facilities;

103

- assessment of health and safety hazards;
- arrangements for work force involvement;
- work permits/systems of work;
- accident and incident reporting;
- emergency procedures.

## CONTRACTOR RESPONSIBILITIES

The contract should make it clear that the contractor is responsible for:

- adhering to the higher of relevant legislation or company policies, procedures and standards or good industrial practice. Specifically, contracts where the company manages or is legally responsible for the operation should require adherence to company policies and standards;
- enforcing of procedures and practices which are safe and without risk to people, property or the environment;
- providing facilities and/or equipment which are well designed and constructed, properly maintained and adequately tested so that the contract can be fulfilled safely and without risk to people, property or the environment;
- reporting all accidents, incidents, injuries and near misses with serious potential. All unsafe acts and conditions should be corrected immediately;
- providing adequate safety and environmental protection policies and regulations. Where necessary a Safety Officer should be appointed to ensure compliance and provide advice;
- providing any necessary safety and protective equipment and ensuring that it is used;
- providing an adequate industrial hygiene programme and compliance with the Control of Substances Hazardous to Health Regulations;
- providing, where necessary, a base safety case and a safety management system which have been accepted by the regulatory authority;
- providing personnel who are medically fit to carry out the work, properly qualified and trained and adequately experienced and supervised.

## COMPANY RESPONSIBILITIES

The company should be responsible for:

- adequately defining the scope of work so that the contractor can properly assess what is required and identify all hazards involved;
- assessing the bids to ensure that all hazards have been recognized by the contractor;

- inspecting, where necessary, the proposed facility and equipment to assess whether it is adequate for the work involved and is able to complete it safely and without risk to people, property and the environment;
- auditing the personnel to be supplied to ensure that there are enough qualified and experienced personnel and supervision to allow the job to be done safely;
- agreeing where appropriate with the contractor whether the company accident, incident and near miss reporting procedure will be utilized or whether the contractor's own system is acceptable as well as how emergencies will be handled and developing and publishing a suitable procedure;
- supporting and encouraging the contractor to achieve a safe operation;
- monitoring the contractor's performance and, if it falls below the acceptable level, taking any necessary action up to and including termination of the contract.

## CONTRACT CLAUSES

The following are possible clauses which could be included in all standard contracts:

- so far as is reasonably practicable, the contractor shall take all precautions necessary to protect the environment, property and his own employees and any employees of the company and other persons who are at any time directly or indirectly affected by the operations of the contractor. Such precautions shall include but in no way be limited to the provision of information on the equipment and substances to be used and hazards involved in the performance of the services or work;
- once on site, the contractor shall familiarize himself with the area of the site where the services or work are to be performed and any operating units bordering the same and the hazards which might be encountered in carrying out the services or work for which he has contracted. The contractor shall co-operate fully with and comply with any directions from the company or any regulatory authority should any of them consider there to be a safety hazard and request the contractor therefore to alter his mode of operations;
- the contractor shall be under an obligation to take all reasonable safety measures in relation to the type of services or work undertaken and shall conduct himself, manage his work force and carry out his operation in such a way as to comply at all times with obligations and duties under the appropriate health and safety legislation, any and all enactments made under authority of that legislation and all other enactments in force from time to time relating to health and safety matters;
- the contractor shall in addition observe and follow all guides, codes and recommendations issued or made by any government, professional or trade

organization or other official or responsible organization relating to health and safety at work;

• the contractor shall report immediately to the company all incidents, accidents, injuries or near misses arising from performance of the services or work, giving full details of the relevant incidents together with such other information that the company may require. The company shall have the right to carry out investigations of such incidents and the contractor shall render such assistance as the company may reasonably require in such investigations;

• the contractor shall ensure compliance with safety regulations commensurate with the nature and extent of the services or work. The contractor shall, if required by the company to do so, appoint a Safety Officer who will be responsible for all personnel engaged under the contract including subcontractors. The Safety Officer will draw up and ensure compliance with safety regulations commensurate with the hazardous nature of the services or work and based on the company safety and health policy manual. The regulations shall include the mandatory wearing of hard hats, safety (steel capped) footwear, weatherproof outerwear, gloves, suitable protection for eyes and ears. Fire retardant outerwear shall be provided for all personnel whose work involves hydrocarbons;

• the contractor shall provide at his own expense all protective clothing and safety equipment required by his personnel to comply with the safety regulations.

• no employees of the contractor or of the contractor's subcontractors or any other persons employed through the contractor shall be assigned to work on the contract unless they are trained to the standards contained in the company safety training policy. Where no standards exist in the company policy, the contractor's standards must be acceptable to the company;

• without prejudice to the contractor's obligations as set out herein, the company shall have the right at any time and from time to time to carry out safety inspections and safety audits during performance of the services hereunder.

## PURCHASE ORDER CLAUSES

The following clauses are sample clauses which could be included in all purchase orders:

• the seller warrants that the goods are in strict compliance with all applicable laws, regulations and requirements relating to safety;

• the seller shall take all reasonable precautions and safety measures necessary to protect his own and the company's employees and any other persons who are at any time directly or indirectly affected by the seller's operations;

• the seller shall provide any necessary information to the company to ensure that the goods can be transported, stored, used and disposed of safely and without

risk to personnel, property or the environment;
- the seller shall comply at all times with all laws and enactments relating to health and safety. The seller shall in addition observe and follow any appropriate guides, codes and recommendations issued or made by any government, professional or trade organization or any other official or responsible organization relating to health and safety at work.

## HIRE AGREEMENT CLAUSES

The following clauses are sample clauses which could be included in hire agreements:
- the supplier warrants that the goods are in strict compliance with all applicable laws, regulations and requirements relating to safety. The supplier shall ensure that the equipment is properly maintained and serviced and inspected and tested as required and complete with fully operational guards and safety devices as appropriate;
- the supplier shall provide any necessary information to the company to ensure that the goods can be transported, stored, used and disposed of safely and without risk to personnel, property or the environment;
- the supplier shall comply at all times with all laws and enactments relating to health and safety and shall in addition observe and follow any appropriate guides, codes and recommendations issued or made by any government, professional or trade organization or any other official or responsible organization relating to health and safety at work.

## PRE-START-UP MEETING

Major contracts may include a pre-start-up meeting. This is to be encouraged where the work is very complex or hazardous or involves a number of different companies. The meeting should be called by the person in the company responsible for the technical administration of the contract, and bring together all company personnel and contractors involved in the various phases of the work. Where additional contractors are brought in after the start-up of the work, it may be necessary to hold additional meetings. The contractor's representative should be the contractor's manager responsible for the contract, although the manager may be accompanied by additional personnel.

The agenda of the meeting should include the following:
- state the company's commitment to safety and health and the principles and standards expected. Review the company's safety, health and environmental protection policy and the policy manual;

- review the contractor's safety policy and discuss any divergencies from company policy;
- review the contractor's safety programme and procedures; resolve any areas of conflict with site policies;
- review relevant sections of the company's training policy and required standards and agree how the contractor will correct any deficiencies. Agree employee orientation;
- review site specific conditions, alarms, protective clothing (including the eye protection policy and use of fire resistant clothing for work involving hydrocarbons) requirements, permit system, etc;
- agree how hazards will be identified, contained and communicated to all personnel. Confirm that proper personnel protection equipment will be provided by the contractor — for example, hat, gloves, boots, coveralls;
- agree what provisions will be made for safety adviser coverage and authority;
- review the company accident, incident and near miss reporting procedure and contractor procedure, agree which system will be used and the detailed procedure to be implemented;
- review the contractor and the company emergency procedures and agree how they will be integrated;
- review the company audit and inspection programme;
- review the contractor's responsibilities for its subcontractors and how the contractor will supervise and monitor their actions and performance;
- review the contractor's industrial hygiene programme and agree how it will be integrated in the company programme;
- review the contractor's safety management system (SMS) and agree how it will integrate with the company SMS;
- agree interfaces and communication routes between the company and contractor's management;
- review the company's work force consultation and involvement policy;
- agree how contractor and subcontractor employees will be made aware of the policy and how it will be applied.

## TRAINING

It is essential for safe operations that the personnel who carry out a job are properly qualified and experienced. This means that a person must have any necessary technical training and safety training and have sufficient experience to be able to carry out the work safely, bearing in mind the quality and quantity of supervision available. Standards of the safety training should be those which are given in the company safety training policy manual. The person defining the technical content of the contract should be responsible for defining the technical training and experience required for the various categories of personnel involved in the contract.

## ON-SITE SAFETY PROGRAMME

The following elements should be included in the on-site safety programme. The list is not exhaustive and it may well be necessary to introduce additional topics to correct deficiencies or contain specific hazards.

### INDOCTRINATION

All new arrivals, no matter how short the stay, should receive an orientation covering emergency actions, routes, the potential hazards on the site, the work permit, isolation and any other procedures which affect them, and also how the job they are to do interacts with the rest of the site activities.

### TOOLBOX MEETINGS

Prior to starting a new task, the leader of the group should call a meeting of all those to be involved to discuss work procedures and safety implications.

### SAFETY MEETINGS

Every person, employee or contractor should attend a safety meeting at least once per month where current safety items are discussed and safety problems analysed. These meetings should be run by the relevant supervisor and give all individuals the opportunity to represent in person any safety matter that concerns them.

### MAJOR TASK BRIEFINGS

Before any major task is undertaken — for example, prior to commencing a major maintenance shutdown — a meeting of all supervisors on site should be held to discuss requirements and review safety interactions and implications. This meeting should be followed by toolbox meetings for all relevant groups.

### SAFETY INSPECTIONS

Every supervisor should carry out an inspection of his area of responsibility at least once per week. The whole site should be inspected monthly. Where possible the inspection team should include someone external to the site and a company representative.

Annual safety audits should be carried out by a management team which contains representatives from the company and contractor's corporate management.

## EMERGENCY PROCEDURES

The company employee responsible for administrating a contract involving the provision of facilities and/or labour must ensure that an adequate emergency procedure is in place and that personnel, contractor and company are adequately trained. Sufficient drills and exercises should be carried out to provide confidence that an emergency will be competently handled. It is essential that the company Safety Manager is involved in the discussions on the emergency procedure to ensure that the contractor procedure dovetails into the company's. This should normally be arranged at the pre-start-up meeting.

Appropriate personnel should be reminded that it is necessary to obtain emergency procedures/actions and recommendations for contracts involving the provision of equipment, since these procedures need to be included in the facility's emergency procedure.

## SUPERVISION, MONITORING AND REPORTING

Although the primary obligation for the safety of a contractor's operation rests with the contractor, the company should ensure that, following the award of any contract, the contractor's performance is commensurate with that of the company's standards and practices as specified in the contract. This should include:
- providing regular supervision of contractors by qualified company staff;
- monitoring the quality, condition and integrity of contractor's plant, equipment and tools;
- carrying out safety audits/inspections of contractor's operations;
- investigating significant contractor accidents/incidents and monitoring follow-up.

## SAFETY SUPERVISION

The company representative on site should supervise the operation to ensure that the contractor is working safely and adhering to agreed practices and procedures. Specifically the representative should:
- verify that all safety-related clauses in the contract are being complied with;
- arrange regular safety review meetings with the contractor (including review of accident/injury reporting, accident investigation and follow-up);
- have the right at any time to visit contractor operations, to review contractor safety procedures and practices and to inspect/audit contractor safety equipment and facilities;
- have the right to curtail or stop contractor activities for safety reasons with full support from company management;

- regularly provide an evaluation of the contractor's safety performance to company management;
- propose measures and help to improve the company's or contractor's safety programme.

To be qualified to execute their duties properly, company supervisors should receive technical and safety training, and should be experienced in local conditions (ethnic, cultural, language, etc).

During some major activities it may be useful to appoint a temporary dedicated company Safety Supervisor, in order to assist in the efficient supervision of the job. The responsibilities of the Safety Supervisor should be clearly defined in the contract, and also known to the contractor's management.

For professional advice on safe practices and safety management procedures, the supervisor should seek guidance from the company's Safety Manager.

Too close supervision, direction or control by the company could be seen as creating a relationship of master and servant with the risk of the company being held responsible or liable for any contractor's failure. This could relieve the contractor of some contractual responsibilities. For this reason, every contract should clearly state that the contractor is engaged in a contract for services and is at all times independent of the company.

## INSPECTION AND AUDITING

Any contractor found to be working unsafely at any time should be advised immediately by the company representative and called upon to take immediate corrective action. If need be, work should be stopped until the situation is rectified.

Inspection and auditing are important means of monitoring contractor safety. An inspection usually consists of an *ad hoc* check by a company supervisor exercising judgement on the performance of the contractor. It is therefore advisable that the contract makes it clear that the contractor is not the servant or agent of the company and remains at all times an independent contractor responsible for its own activities.

Audits comprise more formal investigations against written standards (included in the contract) and are reported to management. The contractor should be encouraged to adopt such inspections and audits to look at its own organization and, if applicable, that of subcontractors.

Unsafe practices will be discouraged by regular inspection at the work site, while the operational condition of safety equipment and the reliability, serviceability and maintenance of work tools and equipment can be established by unannounced spot checks. Such inspections are particularly necessary for the small local contractor. The effectiveness of these inspections will be much enhanced by involving the contractor's responsible staff in the visits.

Safe operating practices essentially require a safety-conscious attitude. Contractor personnel (including the small local contractor) can only be expected to work safely if company employees also display such safety consciousness in their daily work.

Regular auditing of contractor's safety procedures and safe practices should become a routine practice, utilizing appropriate check lists. Guidance and assistance for such auditing can be obtained from the company's Safety Department. It is recommended that the site management take part in the company audit programme.

It is advisable to stipulate in the contract that delays in work as a result of stoppages due to safety deficiencies on the part of the contractor do not constitute *force majeure*. Each serious violation by the contractor should be duly recorded, a written complaint sent, a written report demanded in return and the contractor advised that corrective 'follow-up' action will be monitored. For failure to observe safety standards by the contractor, resort should be made to such remedial powers as may be written into the contract and/or consideration be given to removing the contractor's name from the bid list for future work. The company should have the power to terminate the contract in the case of serious or repeated safety infringements, and this should be clearly specified in the contract.

### INCIDENT AND ACCIDENT REPORTING AND INVESTIGATION

Adequate incident/accident/near miss reporting and investigation is a pre-requisite for effective management of contractor safety. The contract should therefore require the contractor to report all accidents, incidents, injuries and near misses with serious potential. There should also be a similar obligation for contractors to report any incidents involving harm to third parties.

The reporting procedure and report format should be agreed between the company and the contractor. It is preferable for the contractor to submit the standard company report form; however, it is acceptable to use the contractor's own system provided it supplies the same information as the company report and it is agreed at the pre-start-up meeting.

### COMPANY/CONTRACTOR MEETING

The company employee responsible for the technical management of the contract should hold regular safety meetings with the contractor's management. The frequency will depend upon the complexity and level of hazard involved but should be at least monthly. The meeting should preferably include all contractors involved on the site. However, it may be necessary at times to hold meetings with specific contractors. The contractors should be represented by the manager responsible for the contract. The meeting agenda should include these actions:

• review all accidents, incidents, injuries and near misses to identify problem areas or trends. The results of the investigation of the occurrence should be reviewed along with the recommendations to prevent a recurrence and correction of the basic causes. A timetable for implementation should be agreed;

• review all inspection and audit reports and agree a timetable for correction;

• review progress on previous action list;

• review current and future work to identify hazards and interactions and agree precautions and actions to be taken;

• discuss the safety programme and activities and agree new initiatives as appropriate.

## POST-JOB CRITIQUE

At the end of each job a contractor safety performance report should be completed and submitted to the Safety Manager within one month of the completion of the contract, whilst events are still fresh in the mind, to assist in future assessment of tenders.

# 10.  ACCIDENTS AND INCIDENTS

Despite the efforts outlined earlier in this book, any work place will contain some residual hazards which have either not been recognized or are considered sufficiently unlikely to occur that the level of risk has been accepted. The fact that the hazard has not been identified does not necessarily cast doubt on the professionalism of the safety assessment team — after all, it is impossible to visualize every possible combination of natural occurrence, design weakness, operational failure and human error.

Thus things will occasionally go wrong and an incident or emergency will occur. Many people get confused between these two words. An incident is 'an unplanned event which causes or could cause under different circumstances injury or damage to property, product or the environment'. An emergency is 'an incident which results in a continuously increased level of risk of further injury or damage'.

To identify why accidents and emergencies occur, consider the model shown in Figure 10.1. Health and safety policies are turned into working reality by the arrangements put into place. Despite these arrangements some hazards will inevitably occur. It has to be recognized, however, that hazards can exist often for extremely long periods of time without an accident or effect on health. So it seems as if some sort of threshold has to be crossed before an incident occurs (see discussion on hazard causes in Chapter 8).

A hazard occurs when there is an unsafe condition or an unsafe act. An unsafe condition can be defined as an inadequacy in the situation or system of work, whilst an unsafe act is a personal failure for any reason. Table 10.1 on page 116 provides a list of unsafe acts and conditions which together make up the likely immediate causes.

There have to be reasons for unsafe acts and conditions to exist; these are known as the basic or underlying causes. The basic causes are due to failures in:
- personal factors — for example, physical/mental capabilities, stress;
- knowledge, training or skill;
- motivation;
- supervision;
- equipment/facility design or standards;
- maintenance;
- procedures.

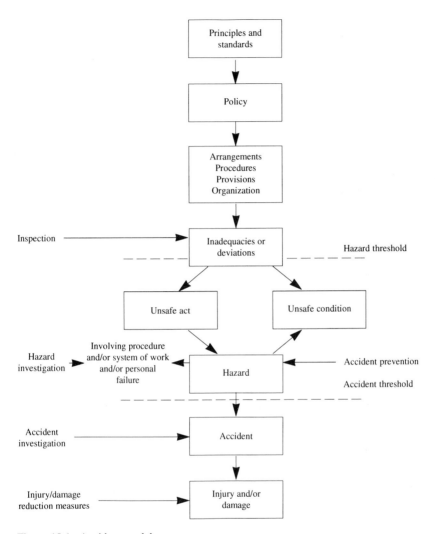

Figure 10.1 Accident model.

The basic causes are due either to inadequacy in arrangements, procedures, provisions or the organization, or to a deviation from these arrangements which are, of course, based on the policy and standards.

## CLASSIFICATION OF INCIDENT POTENTIAL

It has to be accepted that, despite the best efforts to achieve a safe place of work and use safe systems of work, it is very difficult to foresee every eventuality and

**TABLE 10.1**
**Immediate causes**

| | |
|---|---|
| • operating without authority | • failure to use safety devices |
| • failure in communication | • inadequate personal protection equipment |
| • failure to follow rules | • failure to use personal protection equipment |
| • failure to follow procedures | • failure of foundations |
| • inadequate warning devices | • failure of structure |
| • failure to observe warning devices | • inadequate/failure of equipment |

accidents and incidents will occasionally occur. Thus it is necessary to learn from experience. Most companies closely examine all injuries and major incidents which occur to try to establish what weaknesses in their systems allowed the event to happen. But as safety standards rise there will be fewer and fewer accidents to examine, yet there may be a big accident just waiting to happen. In addition there is little point in spending a lot of expensive and precious time in investigating trivial incidents which, no matter what the circumstances, could never be anything bigger. There are one or two techniques which can be used so that a much larger data base of significant events can be examined. These are near miss recording and incident recall.

As many safety professionals know, there are very many occurrences which do not involve injury or significant damage but which provide pointers to hazards which already exist. By close examination the underlying causes can be identified. These occurrences can only be investigated if they are reported. A near miss recording system involves personnel reporting these events as they occur, whereas incident recall is a review of all occurrences involving a piece of equipment or system which a group of workers can recall. Whichever system is being used, a large list of events will be assembled — far more than can possibly be investigated. A system is needed to identify the important ones — to sift the wheat from the chaff. The way to do this is to consider the potential for escalation of each event. If the potential is high then a full investigation should be carried out, whereas if the potential is low it can be ignored. The cut-off point depends upon where the company is in the development of its safety performance. Early on there will be enough serious potential incidents to keep everyone busy, but as the performance rises the threshold should be lowered. It is doubtful if it is ever justified to spend effort on very low potential occurrences. There will always be more useful and productive ways of spending the time and effort involved.

To assess the potential of an event, use the technique for simple risk assessment described in Chapter 5.

## INVESTIGATION

Having decided which occurrences should be investigated to obtain the maximum benefit, it is essential that the investigations are carried out effectively. Accident and incident investigation is like any other form of management decision-making as it involves:

- collecting information;
- checking the information and establishing 'the facts';
- selecting relevant facts to investigate further;
- analysing these facts;
- deciding the most probable causes;
- deciding the most effective actions to take;
- notifying the appropriate personnel;
- recording the events, facts, conclusions, decisions and actions taken;
- checking that the agreed actions are implemented;
- reporting to senior management and the work force.

Collecting the information and facts involves observing the site of the occurrence, and any other relevant area or event, and interviewing all available personnel involved either directly or on the periphery. The visit to the site should be made as soon after the event as it is possible and safe to do. Whilst it will frequently not be appropriate to interview injured personnel at the time, as many witnesses as possible should be interviewed whilst their memory is still fresh and before they start to rationalize and modify their recollections. It is best to interview the witnesses at the scene; it is easier for them to demonstrate what happened than to try to visualize the scene as well as the events. Interviewing is probably the most important skill required for effective accident investigation as it frequently provides the most detailed information and even occasionally the only information. The basic rules for successful interviewing are:

- prepare thoroughly;
- establish rapport and put the interviewee at ease;
- use open questions (which cannot be answered with a simple 'yes' or 'no');
- avoid closed questions (which can be answered with 'yes' or 'no'), and leading questions;
- do not express an opinion or argue;
- take enough notes to refresh the memory afterwards but do not take verbatim notes;
- keep an open mind.

Check the facts as much as possible and where appropriate carry out a further investigation. Arrange and analyse the facts. Establish:
- who;
- what;
- where;
- when;
- the size.

The facts are then analysed in three ways:
- who or what was involved;
- what hazards were present;
- what controls failed.

'Who or what' can be people, property, process and/or the environment. 'What hazards' can be chemical, physical or biological. 'What controls failed' can be barriers, personal protection, protection equipment, procedural or management controls. The analysis should consider the standards set and any deviations from them or changes made to them.

From this analysis, possibly supplemented by some additional collection of information, it should be possible to establish the most likely cause. It is sometimes useful to consider other possible causes as well. Wherever possible the most likely cause should be tested by identifying some other change or deviation which would have occurred at the same time and checking whether it did actually occur. A useful aid to this process is the International Loss Control Institute (ILCI) Systematic Cause Analysis Technique (SCAT chart) which helps to identify the type of contact, the immediate causes and the basic causes.

The next step is to identify the actions which will be taken to prevent a recurrence of the event. This is done using traditional problem-solving techniques:
- define what you are trying to achieve;
- specify the characteristics of the ideal solution;
- divide these characteristics into 'musts', 'importants', 'significants', 'minors' and 'unimportants';
- brainstorm all possible recommendations;
- assess the effectiveness of each possible recommendation against the characteristics.

Again ILCI has a useful aid to generating possible recommendations called the CAN (Control Actions Needed) chart which lists many possible actions to prevent a recurrence by either introducing a new programme or upgrading existing standards. It is also sometimes useful to produce a failure tree analysis which depicts the things which went wrong using the topics list in Table 10.2. Its use is similar to that of standard fault tree analysis to depict what could go wrong to cause a specific top event.

**TABLE 10.2**
**Failure tree analysis topics**

| | |
|---|---|
| MANAGEMENT | Supervision, motivation and priorities |
| MAN | Recruitment standard, skill, personal factors and knowledge |
| MACHINE | Equipment, tools, design and maintenance |
| MATERIALS | Substances and protective equipment |
| METHODS | Work standards and abuse/misuse/erroneous procedures |
| ENVIRONMENT | Physical and climatic |

**TRAINING**

It is probably obvious that it is essential to train supervisors and managers how to carry out effective investigations. The training session should include the following topics:
- the effects and costs involved in accidents;
- types of accidents and the accident triangle (see Figure 8.2 on page 86);
- how accidents happen, hazards, immediate causes and basic causes (see Chapter 8);
- site investigation;
- interviewing;
- decision-making;
- reporting.

Case studies and role play are useful techniques to assist in effective training. A number of packages are now available commercially.

**REPORTING**

All accidents and near misses must be reported and an analysis carried out to identify those with high potential which warrant full investigation. Thus a simple one-page report form should be filled in by the observer or first aider involved in any occurrence saying who, where and when, classifying the outcome and describing the event briefly. The form should be passed to the relevant supervisor who assesses the potential exposure and consequences to determine the level of risk (as discussed in Chapter 8). Where the risk level is assessed as medium or higher, a full investigation should be carried out. The form should then be copied to the appropriate Manager and the Safety Department. The Manager, in conjunction with the Safety Manager, should review all very high potential

119

occurrences to determine whether a more detailed investigation is necessary to identify any company-wide implications and potential. Such an investigation should describe the details of any injuries and damage, listing the equipment involved. A suitable classification of injuries is the UK Health and Safety Executive Accident Prevention Advisory Unit (APAU) classification. The occurrence itself must be described, including the operations under way at the time, the location and the agent which caused the incident (again the APAU provides a useful classification). The most important part of the form then follows, giving the unsafe acts and conditions (the immediate causes) and the basic or underlying causes. These must be followed by a list of the actions which have already be taken to prevent a recurrence, plus any planned actions with target dates for their implementation. The form must have space for comments by the Manager and Safety Manager.

Having reported the occurrence and investigated it, it is essential that there is an adequate system to follow up and track the recommendations to ensure they are implemented in a timely manner and to allow regular progress reports to be submitted to senior management.


## STATISTICAL ANALYSIS

The Safety Department should generate an analysis of all injuries, incidents and near misses regularly, at least once a year but more often if there are enough occurrences to make the analysis meaningful. The analysis should compare the actual number of each type of incident against the goal for the relevant period and past experience. This will indicate whether the trend is upwards or downwards. In addition the frequency rate for the various classes of incident — for example, fatality, major injury, reportable injury, minor injury and damage — should be compared with previous years, other parts of the company, competitors and the overall industry average if these are available, to get an indication of whether the performance is improving or not and how it compares with the rest of the company and the industry average performance. Once performance has improved, the numbers will be very low and individual events will cause a major change. In these circumstances it is better to use a three-monthly moving average rather than individual monthly figures.

The investigation reports should also be analysed by all the parameters available — for example, department, type of operation under way, part of body injured, nature of injury, agent causing the injury and the incident type. Other interesting statistics can be obtained — such as how many incidents happen on each day of the week, each hour of the day, or the length of time the person worked before the incident occurred, how much experience of the job the person had, etc. Analyses such as these can throw a lot of light on problem areas and allow attention to be directed to the areas most likely to bring improvements in performance.

The one-page initial report should not be discarded since analysis of it can also indicate potentially fruitful areas for further investigation. A sudden rash of similar incidents can indicate a developing problem or the effects of some change in the system. An easy way to assemble this data is to set up a matrix either manually or on a computer. The matrix is an alphanumeric reporting system with an infinite reporting capability by inserting an 'x' for each item deemed relevant in each individual incident. The matrix uses letters across the top and numbers vertically. The letters identify groups of information whilst the numbers subdivide the groups into elements. Minor incidents for which a superficial investigation will have been carried out only allow partial completion, but this is adequate for a coarse analysis.

By adding up the number of 'xs' in each box and comparing it with the total number of entries, problem areas are identified. These areas warrant a more detailed investigation to see if there are common unsafe acts or conditions or basic causes. The alternative to this system is a detailed investigation of every event and a complicated recording system to identify common areas.

Accident and incident investigation identifies useful information about the hardware and software deficiencies. However, only limited information about human failures will be developed unless a specific human error identification study is carried out. People do not willingly volunteer information about their errors. Because of this, near miss reports mainly concentrate on equipment and procedural failures. One of the few techniques available which allows fuller information to be obtained on human errors is the use of a confidential reporting system as operated by the aviation industry. This scheme encourages personnel to report personal mistakes and concerns freely and confidentially to an independent body without fear of retribution. The lessons learned are then passed on to the whole industry allowing improvements and adjustments to be made before a major accident occurs.

Having gained as much knowledge and experience as possible from every unwanted event, it is essential to ensure that this information is retained within the company. Otherwise in a few years the lessons will have been forgotten, similar incidents will occur and the same learning curve will be repeated. The only practical way to avoid re-inventing the wheel is to build the lessons learned into the company standards and procedures so that they will not be forgotten.

# 11.   EMERGENCIES

The previous chapter discussed how incidents occur and reviewed how to investigate and report them. In many industries, however, incidents can very easily turn into emergencies and there is no sharp dividing line between the two. The following definitions, already used in Chapter 10, differentiate between the two and explain some other terms used during emergencies:

• incident — an unplanned event which causes or could cause under different circumstances injury, ill health or damage to property, or the environment;

• emergency — a dynamic incident in which there is a continuing potential for major injury, ill health or damage to property, the process or the environment;

• evacuation — the planned and controlled removal of personnel from an emergency area;

• escape — the uncontrolled departure of personnel from an emergency area;

• rescue — the recovery of personnel from an area of danger to a safe location.

An emergency normally develops as shown in Figure 11.1. At any time during normal operations an upset can occur because of a change in a process parameter, the failure of a piece of equipment or instrument or control system, or an error by a human being. Minor process deviations will usually be corrected by the plant control system without any involvement by the operators.

Larger deviations normally require the operator to take control, correct the situation and stabilize the system again. But some deviations will be so large or impossible to correct that it will be necessary to shut the plant down. Generally this occurs automatically if the process parameters go outside the operating envelope; however, it is sometimes appropriate for the operators to carry out a controlled shutdown.

These shutdowns are designed to allow the early start-up of the plant again. If the situation has deteriorated to the point of loss of control and the inability to regain it, an emergency shutdown must be initiated in which the various sections of the plant are isolated and the hazards eliminated or contained. At this stage an emergency is usually declared and the emergency procedures instituted until someone in authority declares the situation to be safe again. It is normal to sound an alarm, muster all personnel for a head count and locate any missing personnel.

If the situation looks likely to deteriorate further, an evacuation will be ordered. It must be recognized that at any time the situation can deteriorate so

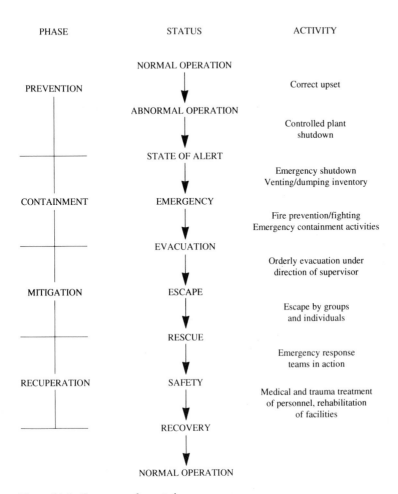

| PHASE | STATUS | ACTIVITY |
|---|---|---|

NORMAL OPERATION

PREVENTION — Correct upset

ABNORMAL OPERATION

Controlled plant
shutdown

STATE OF ALERT

Emergency shutdown
Venting/dumping inventory

CONTAINMENT — EMERGENCY

Fire prevention/fighting
Emergency containment activities

EVACUATION

Orderly evacuation under
direction of supervisor

MITIGATION — ESCAPE

Escape by groups
and individuals

RESCUE

Emergency response
teams in action

RECUPERATION — SAFETY

Medical and trauma treatment
of personnel, rehabilitation
of facilities

RECOVERY

NORMAL OPERATION

Figure 11.1  Sequence of events in an emergency.

rapidly that it is not possible for this sequence to take place, or it may be tele-scoped so much that the corrective actions are ineffective and personnel must make their own decisions to escape either in groups or individually. Personnel may become trapped and need to be rescued. In addition, depending upon the type of industry and the location of the facility, personnel who have been evacu-ated or escaped may not be in a truly safe location and they will also need to be rescued at some stage.

Survivors from a major emergency suffer very high levels of stress and trauma and will need appropriate treatment. In addition, once the emergency is over, the facility must be rehabilitated before normal operations can be restarted.

Figure 11.2 shows the actions and decisions which a company should take to prevent and mitigate the consequences of emergencies. Once the basis of design of a facility has been fixed and the safety assessment carried out as discussed in Chapter 5, a number of residual accidental events will remain in the

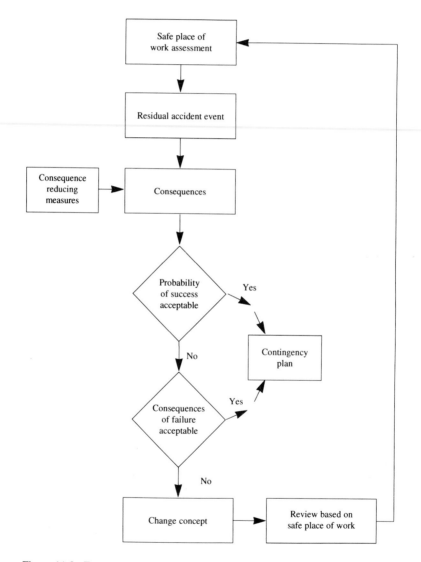

Figure 11.2   Emergency response decisions.

design. The consequences of these events should be analysed along with the effects of any mitigating measures — for example, fixed fire fighting facilities — to decide whether the mitigating measures will be successful. If a successful outcome is unlikely then a decision must be made about whether the resultant consequences are acceptable or not. If they are acceptable, an emergency plan must be generated detailing the facilities to be provided and actions to be taken to protect personnel on and off site. If the consequences are unacceptable, the basis for design must be changed so that the design residual events are altered until the consequences are acceptable.

## TYPES OF EMERGENCY

The preparations for dealing with emergencies must take into account and be able to cope with a wide variety of emergencies including, where relevant:
- internal operational incident — fire, explosion and toxic release;
- external threat — fire, explosion or toxic release in an adjacent facility or a transport accident;
- natural disaster — flood, wind, lightning or earthquake;
- civil disorder — riot, demonstration, extortion or threat;
- malicious damage — sabotage or arson.

## EMERGENCY MANAGEMENT

The goals of emergency management can be summarized as to:
- prevent injuries both on and off the plant;
- reduce property damage and maintain the security of company property;
- reduce damage to the environment;
- minimize production downtime;
- maintain good public relations;
- support shareholder confidence and share value.

Emergency management is based on the generation of a satisfactory and adequate set of contingency plans. These operate at a series of different levels depending on the complexity of the organization. As Figure 11.3 on page 126 shows, the facility itself requires an emergency procedure designed to protect personnel on and off site, contain the emergency and minimize damage to the process, property and the environment. At a higher level, managers have a responsibility to provide assistance to the facility and liaise with official agencies. At this level or higher there is also a need to communicate with relatives of employees and contractors, partners in the venture and the media. At corporate or company level information must be supplied to the government and shareholders.

125

LEVEL OF RESPONSE

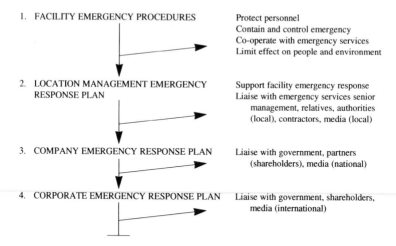

1. FACILITY EMERGENCY PROCEDURES

Protect personnel
Contain and control emergency
Co-operate with emergency services
Limit effect on people and environment

2. LOCATION MANAGEMENT EMERGENCY
   RESPONSE PLAN

Support facility emergency response
Liaise with emergency services senior
   management, relatives, authorities
   (local), contractors, media (local)

3. COMPANY EMERGENCY RESPONSE PLAN

Liaise with government, partners
   (shareholders), media (national)

4. CORPORATE EMERGENCY RESPONSE PLAN

Liaise with government, shareholders,
   media (international)

Figure 11.3   Levels of contingency response.

## SEVERITY OF EMERGENCIES

The level of response is obviously determined by the severity of the emergency. Normally three different categories of emergency are recognized:

CATEGORY 1

A Category 1 emergency may be classified as a situation which does not threaten the safety of the plant itself and has not caused any major injuries to personnel. It can be handled by the plant personnel without any external assistance and it will not generate any significant media attention. Examples are:
- medical assistance to sick or injured personnel;
- a small fire or minor spill which may generate minor, purely local, media attention.

CATEGORY 2

A Category 2 emergency may require some external assistance but will not extend beyond the boundary of the facility. It is likely to generate local media attention and may be picked up by provincial or national media. Examples are:
- multiple injuries;
- a single fatality;

126

- a significant fire;
- a minor explosion.

CATEGORY 3
A Category 3 emergency will put the facility under serious threat and is likely to extend beyond the boundaries or cause multiple injuries or death. It will require assistance from external sources and is likely to generate national media attention.

## FACILITY EMERGENCY PROCEDURES
In brief the facility personnel will try to:
- protect the safety of personnel;
- account for all personnel;
- rescue any casualties and provide medical treatment;
- identify the real nature of the emergency and its likely effects;
- limit as much as possible the damage to property and the environment;
- shut down operations as necessary;
- inform senior personnel;
- liaise with emergency services response teams.
  The emergency procedure should cover:
- responsibilities of company and emergency services personnel;
- organization and co-ordination of emergency control teams;
- reporting requirements and arrangements;
- action plans for every member of the emergency teams;
- procedures for specific types of emergency giving key actions to be considered;
- contact details for emergency services and company personnel;
- emergency equipment lists;
- standard forms to be used during emergency activities.

## FACILITY MANAGEMENT RESPONSE PLANS
In brief the management response will involve:
- mobilizing external resources to support the emergency control team;
- providing technical advice to the emergency control team;
- liaising with the emergency services;
- liaising with contractor management;
- keeping company senior staff informed of developments.
  The response plan will include:
- the response organization;

127

- the reporting requirements;
- the responsibilities of individual response team members;
- the responsibilities of the emergency services and their activities;
- mobilization and callout arrangements;
- check lists of key actions for individual team members;
- facilities available and their location;
- contact details for company personnel, emergency services, governmental and official agencies and sources of assistance;
- *pro formas* covering notification of emergencies, communications, official report forms and press holding statements.


## CORPORATE AND COMPANY EMERGENCY RESPONSE PLANS

The following responsibilities may be combined or split depending upon the size and complexity of the organization:

- ensure that the facility has adequate resources;
- keep the board of directors informed;
- keep senior civil servants and the government informed;
- approve all press statements and take part in press conferences;
- keep partners informed;
- liaise with relatives and next of kin;
- respond to media attention.

The response plan is structured in a similar manner to the facility management response plan.


### PHYSICAL PREPARATIONS

PROCESS HAZARDS MANAGEMENT

A periodic review of all operations involving hazardous materials or processes can significantly reduce the chance of an emergency occurring. The review ensures that all reasonable and practical precautions are being taken. It also helps to ensure that unauthorized changes have not been implemented and that drawings and procedures are up to date. The review should cover:

- operating procedures;
- maintenance procedures;
- modification;
- control of non-routine operations;
- fire precautions;
- process design safety;

- hazard assessment;
- training;
- incidents and their reporting and investigation;
- drawings.

## FIRE PROTECTION PROGRAMME
Fire prevention and protection standards should be kept up to date and be understood by all employees. Supervision and employees should be fully aware of the fire and explosion hazards and protection requirement for their area. Equipment must be adequate and well maintained. Ignition sources must be eliminated or rigorously controlled. Safe practices must be followed in the handling and storing of flammable liquids and gases.

## TOXIC MATERIALS PROTECTION
Since an uncontrolled release of toxic material — whether in the initial incident or during the emergency — will cause a major hazard, it is essential to have available means to isolate the source, to assess the extent of the hazard and to respond to the emergency. Fast response is critical to the safety of personnel both on and off site. The best method of preventing a toxic release emergency is to minimize the inventory, and to utilize a high standard of system integrity, good maintenance and operating procedures.

## EMERGENCY ALARMS
Distinctive alarms are required to warn personnel of an impending emergency. They must be effective in alerting all personnel, with no dead spots — this usually means audible alarms with visual alarms in high noise areas. The emergency alarm system should allow personnel to distinguish between fire and toxic emergencies.

## EMERGENCY SHUTDOWN FACILITIES
A fast-acting emergency shutdown which shuts down and effectively isolates the various segments of the plant and dumps any hazardous inventories to safe locations significantly reduces the chance of a serious emergency developing.

## EMERGENCY RESPONSE CENTRE
Obviously the response centre requires adequate space and furniture for all the members of the response team. Adequate communications and facilities are also essential to the effective functioning of the team. These include:
- unlisted direct exchange telephone lines;
- company exchange telephone lines;

- telex;
- fax facilities;
- possibly radio facilities;
- electronic mail terminal, if used by the company.

The centre should be fitted with white boards for recording a log of events, personnel status, weather conditions and the official agencies contacted. A full range of maps, manuals, drawings, process descriptions, name and address lists and stationery is required. The response centre should be located remote from potential hazards and provide adequate safety to personnel. It should be constructed of non-combustible materials and be blast-resistant if flammable gases or low boiling flammable liquids are handled on site. Since the nature of emergencies is unpredictable, an alternative response centre should be available with sufficient facilities to allow it to be used if the primary centre becomes untenable for any reason.

## EVACUATION

The planning for evacuation is important in preventing injury during a traumatic event. The planning should include:

- who will authorize an evacuation;
- how the decision will be communicated to all personnel;
- what exits will be used;
- what equipment will be available to aid evacuation;
- what safe havens will be used prior to the evacuation;
- how the head count will be taken after the evacuation;
- what assembly points will be used after the evacuation;
- what facilities will be used by the emergency control team after the evacuation;
- how search and rescue will be carried out after the evacuation;
- whether evacuation of the public will be necessary;
- if so, how it will be organized and carried out.

## PUBLIC RELATIONS

A site emergency is not a private affair — fire, smoke and fumes or an explosion will quickly alert the community. Not only will employees' and contractors' relatives be concerned but local authorities, emergency services, government agencies, neighbours and the media will all rapidly be aware of the occurrence and expect information.

Depending on the size of the incident and the phase of the emergency at which the outside world becomes aware of it, handling telephone calls may become a major problem for the emergency team. The only sensible thing to do is to ensure that handling calls from relatives and the media is a controlled activity — which means that it is planned.

By establishing teams to handle these calls from an early point in the response to an emergency, the company takes a major step towards ensuring that all releases of information to third parties are controlled activities and carried out in a structured way. This requires that the emergency control, response, press and relatives teams are all working in full co-operation. It is counter productive to tell the media more than you are telling the relatives and vice versa. Experience has shown that information is very quickly circulated between third parties and any inconsistencies will rebound on the company concerned.

*Relatives response team*

During an emergency the relatives response team will be confronted with a variety of inquiries for information regarding the incident, both in terms of the nature of the request and the credentials of the person making the request. It can come from:
- a close relative or friend;
- a social worker, clergyman, MP or doctor acting on behalf of a relative;
- a company employee on the plant;
- a company employee from another site;
- a contractor working for the company;
- a completely spurious source.

It is possible that a large number of calls will be received that do not relate to the emergency in any way. It is therefore essential to establish very quickly whether a call is relevant and to put to rest all those that are not.

The style of inquiry will vary widely. Very few will be calm and understanding, and the relatives response team must be trained and prepared to cope. The purpose of the relatives response team is to:
- remove pressure from the operational personnel, allowing them to concentrate on controlling the emergency;
- control the release of information to ensure that it is accurate and limit rumours as much as possible;
- gather information on relatives and next of kin to minimize delay in reaching them with essential information as it becomes available;
- avoid unnecessary anguish and worry for relatives;
- avoid damage to the company reputation and image.

*Media response team*

The objectives of the media response team are to:
- remove pressure from the operational emergency control room;

- control the release of information to the media;
- collect information from the callers;
- avoid damage to the company reputation and image.

The role of the team is to:

- provide a service to answer calls from journalists and other members of the media;
- give out approved press statements over the telephone;
- answer over the telephone legitimate questions from the media which are within their sphere of knowledge and which relate to approved press statements;
- collect information from callers concerning questions, rumours, comments, etc, which may require responses from the company.

It is essential that the company is seen as the primary authoritative source of information and that everyone speaking on behalf of the company 'sings from the same hymn sheet'. It is wise to start to issue background information about the organization as soon as possible after the onset of an incident, thus demonstrating a preparedness to communicate during the crisis. This will provide a valuable breathing space to prepare further accurate press statements.

During an emergency it is inevitable that groups outside the company will be contacted by the media for information. It is essential that the company retains control of the information released in order to ensure that such information protects company interests. It is therefore necessary to send a copy of all press releases to the emergency services and other external agencies as soon as they are approved, as well as sending them to the news agencies.

The external agencies will also be generating press releases and these must be reviewed for consistency with the company statements. If reports are not forthcoming from the external agencies, they should be actively sought so that co-ordination and consistency is maintained.

Remember that the media response team requires a source of facts on the company, its background and facilities. It is wise to have a set of 'fast facts' prepared in an easily useable form for use by the team during any emergency.

*Community relations*

Good community relations are important. An emergency which is correctly handled will enhance the company image in the eyes of the local community, whilst a poorly handled incident will ruin months of hard work. An emergency quickly involves a number of community agencies, civil authorities and governmental officials. If it is contained within the company's premises and does not threaten the outside community then the civil authorities have only a limited role in dealing with the emergency. However, if the emergency threatens people or property outside the company boundary they are involved and in many cases take overall control. It is therefore essential that a full understanding of their roles, responsibilities and authorities is established, along with a good working relationship before an emergency occurs.

## POST-INCIDENT RECOVERY

The key to controlling and reducing the cost of an emergency is prompt action following a prearranged plan. Recovery should be started as soon as there is anything to salvage and it is safe to do so. Nothing is gained by having personnel standing by until the emergency is totally over. Stock, equipment and records must be rescued as soon as it is safe to so do to prevent further damage and loss. Property should be restored to as near normal condition as soon as possible.

The first thing to do to achieve a fast recovery is to appoint someone responsible for preparing and implementing a plan for the restoration of operations. Obviously this plan includes a list of readily accessible personnel who can respond as soon as they are notified of an emergency. The list should not include personnel who will be involved in controlling the emergency or the back-up response teams. It will consist of relevant company personnel, contractors who can assist in expediting salvage, repairs and rebuilding, plus suppliers of spare parts for damaged equipment.

The company employees are probably best qualified to carry out salvage since they know the equipment and products and are the best people to inspect it visually. The inspection is aimed at recovering all product and machinery that will be cheaper to repair than replace. A visual inspection is only a guide and it is frequently necessary for a more detailed inspection and/or testing to be carried out to confirm that the product is still up to standard and equipment is useable. What at first sight appears to be a total write-off may not be, and it is often economic to recover the material. Prompt salvage will also prevent further deterioration, which is normally progressive and frequently exponential.

Effective salvage starts before the incident — in the planning which takes place. This should simplify the job of salvage and make it possible to carry out the operation as early as is practical.

Measures such as waterproofing floors, drains which can cope with deluge and fire fighting activities, access to tarpaulins, storing items on pallets to lift them off the floor and ease their removal in an emergency are a few of the actions which can be taken. The capability to remove flammable liquid from areas of risk, or at least isolate it so that it does not contribute to any fire, are important steps. The provision of adequate emergency detection facilities allowing faster detection, shutdown and emergency response minimizes damage. It is of course essential to ensure that these systems are regularly tested to be certain they will operate when required. It is amazing how many serious fires could have been controlled whilst they were small and insignificant, had the automatic fire systems been operational.

Salvage efforts should concentrate on:

- removing stock and equipment from the area;
- protecting removed material and equipment from further damage;
- minimizing the effects of the emergency on areas not directly affected — for example, by ensuring that drains do not become blocked.

It will be necessary to source salvage equipment very quickly — for example, mops, buckets, squeegees, pumps and test gear. This should all be pre-planned. A useful source of help in salvaging equipment is the manufacturer, who has probably done it before.

Once the emergency is over:

- restore emergency protective systems as soon as possible, as the effects of a second emergency are usually far more severe than the first;
- concentrate on caring for the more valuable stock and equipment;
- preserve business continuity;
- keep customers supplied.

## TRAINING, DRILLS AND EXERCISES

Emergency facilities and procedures are of little use if personnel are not adequately trained and competent to carry out the defined actions. Thus emergency training, drills and exercises are designed to ensure that all personnel are prepared to cope with any emergency which arises and ready to carry out any duties assigned to them. Preparations for emergencies must include:

- training — to ensure understanding of the nature of the potential hazards and the facilities, equipment and procedures to be used to handle any emergency which occurs;
- drills — to provide practical training on specific emergency equipment, means of evacuation and escape and the procedures personnel should follow in an emergency. Drills also establish a routine so that personnel are more likely to follow the established procedures in the stress of a real emergency;
- exercises — to demonstrate that personnel are able to respond effectively to an emergency, to identify the strengths and weaknesses in the emergency procedures and any training needs not yet fulfilled.

## TRAINING

Training needs will vary according to:

- the level of risk;
- the applicable legislation and company policies;
- the company organization;
- the emergency procedures;
- the duties the person has to carry out in an emergency;
- the equipment to be used in an emergency;
- the individual's abilities, experience and previous training;
- the amount of direct supervision available.

Thus training must be provided to those with:

- command duties in an emergency;
- co-ordinating and management duties in an emergency;
- specialist duties in an emergency.

In addition it is essential to ensure that all the remaining personnel know what they must do in an emergency. Make arrangements to ensure that any visitors to the site either know what they must do in the event of an alarm, or they must be escorted at all times during their visit.

DRILLS

The overall purpose of a drill is to:

- maintain the competence of personnel to use the emergency equipment they may reasonably be expected to operate in an emergency situation;
- practise the emergency procedures and communications;
- confirm that the emergency equipment is ready to be used at all times.

A plan should be prepared for every drill showing the objective and the means of achieving it. The plan should define the duties and actions expected of everyone involved and these should be checked during the drill.

*Musters*

The purpose of muster drills is to familiarize all personnel with the routes to and the location of their muster point, the alarm signal calling a muster and the head count procedure. All personnel should participate; if it is necessary to excuse essential personnel they must take part in the next muster drill. Where alternative muster points are designated, drills should periodically use these alternative points.

*Evacuation*

Evacuation drills are designed to familiarize all personnel with the evacuation routes and facilities. All personnel should take part, and any essential personnel excused from one drill must take part in the next one.

## Fire fighting

Fire fighting drills are intended for personnel with specific fire fighting duties in the event of an emergency involving fire. The drill should cover the fire team assembly point, use of portable and fixed fire fighting and other emergency equipment and its storage locations including protective equipment, means of tackling different emergency scenarios and means of access to the various parts of the facility. It is essential that during these drills adequate resources are available to deal with any real emergency that arises.

## Breathing apparatus

Personnel who may need to wear or control the use of breathing apparatus (BA) in an emergency should take part in drills to maintain their readiness. Areas to be covered include donning and use, preferably under 'hard work' conditions, practise of the entrapped procedure including conservation of air supply, entry procedure including the tally board, search and rescue procedures and communications whilst wearing BA.

## Emergency equipment

The purpose of emergency equipment drills is to maintain the competence of the emergency team involved in the use of rescue and other specialized emergency equipment such as rescue winches and hydraulic jacks. Such drills include the use of all types of gas detectors available, rescue equipment and other emergency equipment, and the checking of the serviceability of the equipment.

## Casualty handling

Personnel involved in the rescue, evacuation and handling of casualties should take part in regular drills covering use of stretchers, location and use of first aid equipment, use of resuscitation equipment and the care of injured personnel.

## First aid

Qualified first aiders should carry out periodic drills to ensure that they are always ready to administer treatment to injured personnel. The drills should include all aspects of first aid included in the qualification course.

## Drill requirements

Once all the required drills have been identified, generate a table showing the drills, their frequency and who should take part. It can also indicate which drills can be combined with which other drills. Obviously there must be a system for recording drills carried out and who took part in them.

EXERCISES

Test the effectiveness of emergency procedures, training and drills periodically by a programme of emergency exercises. Because of the diverse nature of emergency procedures, it is not possible to be specific about the detailed contents of exercises, but the following aspects should be considered:

• a scenario should be developed for every exercise defining the problem and parameters being covered;

• all personnel involved should be briefed on the limiting parameters but not the problem;

• personnel not involved should be briefed so that they will not react incorrectly to any inadvertent interfaces with the exercise;

• to reduce the chance of inappropriate actions, all announcements and messages must be prefaced by words similar to 'this is an exercise';

• suitably experienced and briefed personnel should observe all critical phases and locations;

• the progress of the exercise and the prevailing operational conditions must be closely monitored to ensure that safety is not being jeopardized;

• a debriefing session of all personnel involved should be held to identify the lessons that can be learned from the exercise;

• a full report should be prepared describing the exercise and the strengths and weaknesses apparent in the procedures, and recommending any necessary improvements;

• a follow-up report should be issued describing the actions taken to implement the recommendations.

*Facility exercise*

A facility exercise is designed to test the effectiveness of the programme of drills being carried out on a facility. It involves the facility operators and management only.

*Major in-house exercise*

The overall state of readiness of the operating company to deal with a major emergency also needs testing. The exercise should involve all operational and management personnel who have a role in the emergency procedure and management response plans. This type of drill should be given a code name and all personnel should be reminded of the importance of using the code name at all times during this exercise. The exercise should involve contractors and people simulating relatives and next of kin, local and national authorities, the emergency services and the press and public. Personnel should be briefed to act as casualties.

*Major exercise*

A major exercise is a very big undertaking and can only be carried out with the co-operation of the local emergency services and authorities. Obviously a team must be set up to plan the exercise and liaise with all the external bodies involved.

# 12.   AUDITING

Having introduced a programme aimed at achieving a high standard of safety, performance must be measured to confirm that the programme is effective. It is necessary to measure:
- accident performance;
- compliance with legislation and industry and company standards;
- the extent to which hazards have been eliminated or controlled;
- the achievement of policy goals.

It is comparatively easy to compare the accident performance against the industry average. The other three measurements, however, can only be made using audit techniques.


## AUDITS

An audit is the direct verification of an activity using a systematic approach to identify strengths and weaknesses by comparison with standards and practices including legislation. Audits cover both software — such as systems and procedures — and hardware — such as equipment and facilities. The objective of a safety audit is to provide verified feedback to management on a sample of actual practices and/or equipment in an operation, and where appropriate to confirm good practices and standards, verify that the standards and procedures are satisfactory and recommend any changes necessary. The benefits of safety auditing include:
- provision of independent, objective feedback;
- identification of previously unrecognized problems;
- identification of ways of preventing accidents;
- verification that standards and procedures are satisfactory or need improvement;
- reinforcement of good practices;
- highlighting of hazards to all personnel;
- communication of management commitment;
- provision of a stimulus to help the achievement of long term goals;
- demonstration of the effectiveness of the company's safety management to third parties.

## TYPES OF AUDIT

There are three basic types of audit:

- a technical and process audit, which looks at the hardware and its operation and maintenance, and compares it against legislative, industry and company standards;
- a specific hazard audit, which looks at how a specific hazard is eliminated or controlled;
- a management audit, which looks at compliance with company policies and objectives.

The objectives of an audit are to *inspect* relevant parts of the facility and its operation, *assess* the compliance against a standard, *communicate* management's commitment to achieving the highest level of safety, *follow up* the results of the inspection and the conclusions drawn and, finally, to assist in *raising* the standards in line with the company policy and objectives.

TECHNICAL AND PROCESS AUDIT

As the name implies, the technical and process audit looks at the design, construction and operating standards of the facilities to ensure that they meet current legal, company and industry standards and codes. This audit should be carried out on a three to five year cycle and results from an awareness that the production process is continually changing as technology steadily improves. What was considered safe and satisfactory a few years ago may now not be considered prudent. In addition, new techniques for assessing safety are developed as time passes. That is not to say that modern techniques are not utilized on older facilities but that the new techniques tend to be applied progressively to solve specific problems or in response to specific incidents. It is therefore appropriate to stop and take stock of the overall situation at intervals and see how the facilities would be designed and operated if a new facility were to be set up from scratch. Where differences are identified between the existing facilities and the current theoretical design, it is necessary to assess the effect that the differences have on safety and implement improvements in the short term, incorporate them in the long term plan or justify why it is not necessary to make any changes.

The following aspects of the design are normally included in the audit:

- environmental conditions and the facility structures;
- process equipment design;
- operating procedures and operating limits;
- fire systems;
- emergency shutdown and isolation, including cause and effect;
- gas release and blast protection;
- area classification;

- vessel and piping capacities;
- pressure safety valve (PSV) and flare capacities (normal blowdown and fire);
- electrical and lighting systems (normal, emergency and uninterruptable systems);
- ergonomics;
- drains;
- vent systems;
- utilities;
- heating, ventilation and air conditioning (HVAC);
- evacuation and escape facilities.

Specific areas should be allocated to engineers of the appropriate discipline. Where necessary, additional specific expertise should be brought in from consulting engineers. A senior engineer should run the audit, co-ordinating the activities and ensuring that adequate resources are available. Periodic reviews should be carried out to report on progress and the results to date. At the end of the audit, prepare an overall report describing the areas looked at, the problems identified and the projects raised to correct the deficiencies. Of course this report must be backed up by detailed reports covering each specific topic.

A special form of this audit is the post-start-up audit which should be carried out about six months after start-up. This reviews the safety, emergency shutdown (ESD), fire and gas systems to ensure that they are operational and meet the design philosophy in practice.

*Environmental conditions and facility structures*
Review the current predictions for the maximum design environmental conditions (including wind, rain, snow and waves) and confirm that the structure and facilities are adequate to withstand them.

*Process equipment*
The current process flows — such as mass and heat — should be compared with the design conditions. Similarly the process equipment design should be compared to current standards.

*Operating procedures and limits*
The current actual operating procedures and operating limits should be compared to the current written procedures and standards. Review any deviation and decide whether it is acceptable. If it is, the documentation should be changed. If it is not then further training is called for. The review must cover start-up, normal operation, shutdown and emergency shutdown.

## Fire systems

The plant fire safety philosophy should be reviewed and confirmed as still satisfactory by analysing all possible scenarios to establish the risk level and predicted consequences. Then examine the current facilities for fire prevention, detection and fighting to confirm their adequacy to contain all the scenarios.

## Emergency shutdown (ESD) and isolation

The logic which controls the plant emergency shutdown and isolation systems is normally defined in a cause and effect matrix which shows the various control actions implemented by each initiating deviation. It is essential that the matrix is up to date; a survey is needed to confirm this. Any changes must be reviewed to confirm that they are essential and do not have any unforseen effects. Finally, test the actual hardware to confirm that it operates as designed and that response times are still satisfactory.

## Gas release and blast protection/toxic effects

All potential sources of gas release should be identified, along with the predicted rate of release, gas cloud size and the blast overpressure or toxic contours. Assess the likely consequences of the predicted blasts and toxic gas effects and check the vulnerability of the emergency systems to identify any necessary improvements. The gas detection system should also be reviewed and tested to confirm its adequacy and operability.

## Area classification

An examination should be carried out to determine what changes have taken place which will influence the area zoning with respect to electrical classification.

## Vessel and piping capacities

This is an examination of the adequacy of the vessels for current and future requirements. Some piping velocities may cause erosion: it is therefore essential to identify any possible erosion problems and implement corrective actions or at least a wall thickness monitoring programme.

## Electrical and lighting systems

The adequacy of the electrical power supplies — normal, emergency and uninterruptable, if provided — should be examined to confirm that they are adequate to meet current and future needs. Test the normal and emergency lighting to ensure that they meet current needs and standards.

*Ergonomics*

The term ergonomics relates primarily to instrumentation and the ease of operation of controls for normal and emergency operations. Shortcomings in the man-machine interfaces should be investigated for improvement.

*Drains*

Many incidents with severe potential have occurred because process fluids get to the wrong place via the drain systems. Thus it is essential that regular thorough checks be carried out on all drain systems to find any previously unrecognized cross-connections or abnormal paths.

*Vents*

Considerations similar to those described for the drain systems apply.

*Utilities*

It is essential that there are no direct uncontrolled connections between the process and the utilities systems.

*Heating, ventilation and air conditioning (HVAC)*

The safety of many facilities depends on the HVAC system. It is therefore critical to ensure that the design conditions are maintained and not changed by someone trying to improve the conditions at their work area — for example, by blocking off an air inlet grille. If there is a real problem it must be properly engineered to maintain the system design criteria and the plant safety.

*Evacuation and escape facilities*

The emergency detection and alarm systems, along with the evacuation and escape facilities and procedures, should be checked to confirm that they are adequate to cope with the current emergency scenarios.

SPECIFIC HAZARD AUDITS

Specific hazard audits are in fact a series of audits carried out annually to ensure that legislative and company requirements are being met. Typical subjects include:

- chemicals;
- radiation;
- crane operations;
- explosives;
- fork-lift trucks;
- equipment guarding;

- gas cylinders;
- training and competency;
- emergency response;
- safety programme.

### Chemical audit

A full inventory of all hazardous substances should be prepared by physically inspecting the whole facility. Having listed the substances, carry out a check to ensure that an up-to-date material safety data sheet is available for each of them. The audit should also include confirmation that a risk assessment has been carried out and that the necessary control measures and monitoring activities have been implemented. Finally, a check of the records should be made to ensure that the necessary files are being kept. This audit should be carried out by an industrial hygienist along with a safety advisor who specializes in chemical safety.

### Radiation audit

The company Radiation Protection Advisor accompanied by the facility Radiation Protection Supervisor should confirm that the local rules are valid and up to date. They then carry out a full facility inspection, including the radioactive materials storage facility, a review of working procedures and a check of all the records for registration, authorization and disposal, source leakage checks, controlled or supervised area records, instrument calibrations, dose records and transportation records.

### Cranes

The company crane expert should audit the facility cranes and lifting equipment, checking maintenance and inspection procedures and records, operating procedures, competence and training standards, storage and control procedures. The expert should ensure that a dropped load study has been carried out and updated as necessary.

### Explosives

The company explosives expert should complete an annual check on the transportation, storage and operating procedures and records of explosives kept on the facilities. In addition, review the competence and training records of the personnel authorized to handle explosives.

### Fork-lift trucks

Inspect any fork-lift trucks used in the facility to ensure that they still meet their design specification and that the specification is still valid for the work they are used on. The training and medical standard of all personnel allowed to operate the trucks should be checked against the relevant company policy.

*Equipment guarding*
All mobile and fixed equipment should be inspected to ensure that no ineffective or missing guards are apparent where there is a rotating shaft or 'nip' or 'squeeze' hazard.

*Gas cylinders*
An inspection of the whole site should be carried out to check that all gas cylinders fully meet the company safety standards with respect to the cylinder and its storage and use.

*Training and competency*
The training section should carry out a rolling audit of all personnel (company and contractor), aiming to cover everyone annually. The audit compares the actual training received and standards of experience and competency with the standards laid down in legislation and the company policy. The areas covered include emergency and safety, technical and process and, where appropriate, leadership and man management.

*Emergency response*
An emergency response audit reviews the emergency procedures, emergency facilities, training of personnel and the programme of drills and exercises.

*Safety programme*
Safety programme auditing began as a subject called 'total loss control' and has recently become highly formalized by being developed into programmes such as the ISRS audit, British Safety Council five star audit, the Chase audit and the Sharp audit programmes.

These are very detailed questionnaires/check lists which are used to evaluate the safety programme currently in place and the adherence to it. Many of the audits quantify the responses to allow comparison of the current audit with previous audits of the facility and similar facilities, and also highlight areas of weakness in the programme or compliance with it as well as the areas of strength.

MANAGEMENT AUDIT
Numerous informal management audits take place. Many of them fail because they degenerate into a housekeeping inspection and the results are very negative.

145

The lack of time and the pressure to look at everything means that nothing is looked at in depth and only a very superficial impression of most aspects of the operation can be gleaned. Management ends up with knowledge of some of the failures of the facility personnel and virtually no knowledge of their successes and the good things being achieved. The facility supervisors feel that the audit is unfair as they receive no recognition, merely blame.

The primary objective of a management audit should be to develop a valid assessment of the conditions that exist at the time of the audit for the benefit of both the facility and head office management. The audit should identify the strengths of the current safety and environmental protection programmes just as clearly as the deficiencies. It should not only determine what is not being done, but should recognize and give credit to the programmes which are good and effective. Thus it is necessary to look at a programme in depth. Is there a formal or informal policy and programme? What does it cover in practice? Is it effective? Are the standards satisfactory and results recorded? Are the deficiencies found corrected in an acceptable time scale? What recommendations, if any, should be considered to improve the current situation?

Thus, the investigator needs an in-depth review of policies and procedures, discussion of their application and effectiveness with supervisors and the personnel applying them and, finally, monitoring of the actual application and standard being achieved.

Obviously in the time available it would not be possible to cover every policy and programme. It is therefore recommended that a number be selected and allocated to specific team members. In addition, all team members should be asked to record any unsafe acts or conditions observed during their inspections, note and comment on two aspects of housekeeping and, finally, the overall safety management of the facility.

To demonstrate the commitment of management, the team should be led by a director and be made up of senior corporate managers plus managers from other locations.

Prior to the audit taking place, an audit procedure must be developed and approved. In addition, specific programmes should be allocated to specific team members. Each team needs some pointers to start them off but once they get into a topic their experience as managers will enable them to follow through and investigate areas of weakness and strength.

An opening briefing should be held at which the team co-ordinator describes the technique and a discussion is held on the detailed organization with all the personnel involved. Each group then sets to work and audits the programmes allocated in conjunction with the relevant facility personnel. The groups tour selected parts of the facility appropriate to the programmes being audited. At the end of the audit each group reports to a closing conference on the programme it audited. They describe the appropriate programme, its effectiveness and any recommendations for improvement. Each group then produces a written report which is forwarded to the safety department which prepares an overall summary.

## THE COMPANY SAFETY AUDIT PROGRAMME

The safety audit programme should be co-ordinated with the overall company audit programme to ensure optimum use of resources and minimize overlap. Equally all aspects with any potential impact on safety should be covered by the programme to ensure that there are no major problems requiring deeper investigation. To ensure this the scope of the work must be fully defined. Any sensitive areas should be highlighted so that the team can deal with them in an appropriate manner.

Many operators allow their partners to audit their facilities independently. It has been found, however, that joint audits by the operator and its partners are even more effective because of the interaction between the personnel. The addition of personnel from corporate headquarters and non-partners has also proved to be very beneficial. The facilities, policies and procedures are looked at by totally fresh minds and cross-fertilization of ideas and concepts results. In fact some companies include a regulator inspector. There are also major benefits to be gained from including safety representatives in the management audit team.

External and internal safety audits should be co-ordinated and scheduled to extend the coverage most efficiently, while bearing in mind the essential differences between them. Internal audits are primarily concerned with verifying compliance with the company's own standards and procedures. Verifying the adequacy of management control of safety, if required, may be a difficult role for internal auditors. External audits are most appropriate for verifying the effectiveness of the internal audits and the adequacy of the standards and procedures.

The size of the audit team depends on the size of the audit. The skills represented in the team should be appropriate to the scope of the audit. There may be advantages if auditors investigate and interview in pairs to provide corroboration and challenge of findings. Pairing internal and external auditors optimizes communication and understanding.

## STANDARDS TO USE WHEN AUDITING

Standards to be used could include procedures, guidelines and practices as well as engineering standards and relevant legislation. Ideally there should be standards covering all management, operations and engineering practices against which the company could be assessed. There can be two areas of difficulty for the auditor: non-existent or inadequate standards for software such as operating procedures and management controls, and hardware built to superseded standards.

Operating practices can be audited against the company's own procedures and policies. Auditors should identify where additional procedures are needed, but not necessarily detail them. Even though standards for management controls are often not formalized, there is usually agreement on the desirable actions. Most disagreements are confined to the priorities. Company safety policies and plans can provide some norms against which the management controls can be audited.

As far as plants built to superseded standards are concerned, the criteria should be first legal compliance, second a consistent safety standard within the company and third upgrading where justified. Upgrading to present-day company standards should be recommended when justified. Equally upgrading to industry standards requires the same clear justification. Automatic retroactive upgrading each time a standard is revised is not necessary. Quantitative risk assessment can be useful in assessing the justification for upgrading to newer standards.

## ESSENTIAL ELEMENTS OF AN AUDIT

The method and approach of the audit itself should be well defined, systematic and specifically detailed to ensure adequate coverage and consistency. The use of check lists and questionnaires helps this systematic approach.

The audit leader should be trained and/or experienced in audit organization and technique. The team should contain appropriate expertise, comparable to the expertise of those being audited. Particularly for external audits, the audit team should have sufficient management experience to be able to give constructive and authoritative advice on management systems and approach.

The scope of the audit is normally a facility or activity and its interfaces with management. Within this scope, the terms of reference should be as widely defined as possible, relying on the professionalism of the audit leader to set reasonable boundaries, rather than restricting access to people or facilities. In particular there should be sufficient freedom to include internal audits of other facilities and the progress of follow-up to previous external audits as a measure of the company's internal controls.

After the initial briefing, interviews normally cover all levels, focusing in as problem areas are identified.

The text of the report should be checked in draft form with the specialists within the company to ensure that the facts are correct and that the recommendations are practicable. The final text of the report is then agreed by the team and presented to management. The discussion should clearly differentiate between situations where the deficiencies can be corrected by the existing systems and those where new management initiatives are required. Each recommendation and action should be as specific as the expertise of the team allows.

## SAFETY AUDIT FOLLOW-UP

Implementing the recommendations of the audit is totally the responsibility of the company management. There is no requirement for the company to implement all the recommendations of the audit, but the minimum is that all deviations are treated as variances from standards, and the reasons for not correcting a deviation should be fully explained, documented and authorized by the responsible manager.

The implementation of audit recommendations should be both routinely monitored by the management and reviewed by subsequent audit teams.

# 13.    SAFETY MANAGEMENT SYSTEMS

If all of the aspects of safety discussed so far have been considered and implemented, where appropriate, it may well be thought that all possible requirements
have been covered. But it is very useful to put together a document describing
the safety management system (SMS) since this will show up any gaps or weaknesses in the overall system. The SMS describes how the company organizes
itself, manages safety, interacts with external organizations and finally upgrades
its performance and systems. These components can be subdivided into:

*Internal*
- Company safety principles, standards and policies.
- Organizational structure and responsibilities.
- Personnel qualifications, experience and training.
- Employee involvement and motivation.
- Safe places of work.
- Safe systems of work.
- Operational activities and support services.
- Accident and incident investigation and reporting.
- Emergency response and preparedness.

*External*
- Selection and control of contractors.
- Reporting to and liaison with the regulatory authority.
- Interface with the emergency services.
- Interface with the media and public.

*Upgrading*
- Monitoring and internal auditing.
- Independent auditing.
- Systematic reappraisal and development of the SMS.

To generate an SMS it is necessary to realize that an SMS sits at the
top of a pyramid as shown in Figure 13.1. The SMS is a document which describes all the supporting policies, procedures and systems. An SMS requires a

Figure 13.1  Hierarchy of safety activities.

systematic definition of all the actions necessary to ensure that all operations are carried out safely. This involves planning, organization, execution and maintenance and ensuring that these are carried out in accordance with company policies and any relevant legislation. Thus the SMS controls what must be done, who will do it and how it will be done, defines if it must be controlled by instructions, procedures, drawings or permits, how its accomplishment will be documented, who will verify that the work has been completed as planned and what records will be kept, by whom and for how long. These considerations can be broken down into components.

## PRINCIPLES, STANDARDS AND POLICIES

A successful organization always has clear and unambiguous objectives supported by an effective management which directs and controls the organization. Thus it is necessary to define the company's safety principles or ethics and from these establish the standards of safety performance that the company aims to achieve. The principles and standards are of course supported by the safety policy based on the policy required by the Health and Safety at Work etc Act 1974 requirements (see Chapter 2).

The health and safety policy should be devised and be seen to contribute to the overall business performance, meet the company's legal responsibilities and develop the safety culture and performance. Thus it will be acceptable to and endorsed by employees, shareholders, partners, customers, the authorities and the public at large. The policies must be cost effective and be designed

151

to encourage performance, develop and motivate personnel, protect personnel, the public, property and the environment, reduce the exposure to risk and liabilities and project the desired image of the company.

## ORGANIZATION, STRUCTURE AND RESPONSIBILITIES

The basic cause of the majority of accidents and incidents is the failure to control the system or job by management or supervision, even if the immediate cause is a human or technical failure. Thus the responsibility for ensuring that the standards of health and safety meet the company goals rests with the senior management and this can only be achieved if effective control is exercised over all activities. To do this everyone must understand and *accept* the responsibilities and authority entrusted to them. Thus the SMS must show the management organization of the company and describe the accountabilities and responsibilities of each position from the Board of Directors down to the lowest level in the company (the job descriptions) and how their performance is assessed and reviewed. This description must include the safety and health advisory personnel and organization provided by the company. The purpose of the policy is to ensure:

- control of the organization;
- allocation of responsibilities and authority;
- allocation of adequate resource;
- defined open and clear communications channels;
- definition of the required competence of all personnel;
- clearly defined planning of how safe operations will be achieved;
- implementation of suitable monitoring of the safety performance;
- adequate auditing of the system to confirm that it is satisfactory and indicate where improvements can be made.

## PERSONNEL QUALIFICATIONS, EXPERIENCE AND TRAINING

Many factors influence individual safety performance. It is essential that individuals are matched to the requirements of their job. Each position in an organization, including managers, needs a recruitment profile which describes the criteria for selecting personnel to fill each post. The criteria include qualifications, experience and personal physical and mental abilities. In addition the company requires some form of training policy which defines the training needed for each position. It is important that there is a system to identify changes in the company operation so that further training requirements can be identified and plans made to carry it out (see Chapter 3).

## EMPLOYEE INVOLVEMENT AND MOTIVATION

The safety performance achieved by a company is totally dependent on the individual motivation and actions of each employee — thus employee involvement and motivation are critical to the level of safety performance. Pooling knowledge and experience can generate major advances in safety and risk reduction. Participation increases 'ownership' — a key element in implementing actions and changes — and achieving the goal of safety becoming 'everyone's business' so that nothing is so important that it cannot be done safely. The SMS should therefore include a full description of the policies, procedures and activities — including the committee structure and the terms of reference — used to create employee involvement and commitment and how relevant and comprehensible information is disseminated throughout the work force. The communications policies and facilities used to ensure full free and open discussion of policies, standards, procedures, hazards, levels of risk, performance and the results of monitoring and audits should be described (see Chapter 4).

## SAFE PLACE OF WORK

A safe operation requires a safe place of work and a description of how this is achieved should be included in the SMS (see Chapter 5). This description must include all types of operation carried out by the company and cover all locations where the company operates, including its offices. Topics to be covered include:

- the design and construction of facilities;
- commissioning and operation;
- modification of facilities;
- safe means of access and egress in normal and emergency situations;
- design environmental loads and hazards;
- safety assessment techniques;
- inspection and audit programmes.

## SAFE SYSTEMS OF WORK

No place of work is ever totally free of hazard and it is therefore necessary to define systems of work which allow safe operation (see Chapter 6). The SMS must include a description of the following:

- procedures (operating and maintenance);
- permit-to-work systems;
- isolation standards;
- safety procedures;
- industrial hygiene procedures.

153

Having safe systems of work available is only half the battle; it is essential to ensure that they are followed. The SMS should include a description of how the company monitors adherence to the systems of work and audits them to detect any weaknesses and identify improvements needed.

## ACTIVITIES AND SUPPORT SERVICES

The SMS must apply to all the activities carried out by the company and apply to all phases of a project — that is, from the cradle to the grave — including all production operations and the support services needed to allow these activities to happen. The SMS should include a description of the policies, standards and procedures used to ensure that each of these activities is carried out safely. Include the monitoring of partner operations where appropriate.

## ACCIDENT AND INCIDENT INVESTIGATION AND REPORTING

One important source of ideas for improvements to the safe place and systems of work is the lessons learned from accidents and incidents and, more importantly, near misses. It is essential to identify the unsafe acts and conditions which were the immediate precursors of the occurrence, but it is even more important to identify the basic causes. If these are not corrected then the same event will happen again in a different guise. The SMS therefore includes not only the definition of the various occurrences and levels of investigation but also the follow-up system to ensure the adequacy of the investigation and implementation of the agreed recommendations (see Chapter 10). To achieve this the SMS needs to cover the systems for:

- reporting occurrences;
- ensuring that the agreed recommendations are implemented;
- training of personnel in accident/incident investigation;
- dissemination of the lessons learned to other facilities and the industry as a whole;
- analysing the occurrences to identify trends and problem areas.

## EMERGENCY RESPONSE AND PREPAREDNESS

It is obvious that every work facility requires adequate emergency response provisions in terms of hardware and procedures. In addition the systems used to ensure that personnel are trained and capable of carrying out the emergency procedures and ready to respond to an emergency must be explained, including evacuation, escape and rescue. Full details of the facilities provided should be described and their effectiveness assessed. Since it is essential that the facilities are always available for use, the maintenance, inspection and testing programmes should be described. One of the key aspects of evacuation is human factors — how people behave under the stress involved in an real emergency. The systems and procedures must be able to cope with this behaviour — how this is designed into the facilities should be reviewed in the SMS. Another aspect which should be described is what arrangements are in place and planned to deal with the traumatized survivors, relatives and emergency response teams, both during the emergency and thereafter (see Chapter 11).

## SELECTION AND CONTROL OF CONTRACTORS
Many companies use large numbers of contractor personnel, so a key element in the SMS is how the company controls the selection and activities of its contractors. The company must clearly define the policies and procedures covering the selection of a contractor, and the contract conditions used to ensure the competency and training of the contractor's employees. The safety and health policies and procedures to be utilized during the contract and the methods of supervision and monitoring of the work should be spelled out. It is essential to ensure the full co-operation of the contractor's management and work force to achieve safe operations (see Chapter 9).

## REPORTING TO, AND LIAISON WITH, THE REGULATORY AUTHORITIES
There are certain legal reporting requirements and great benefit to be obtained by tapping into the regulatory authority's expertise and fund of knowledge. So it is sensible to define how the company will report to and interact with the authorities to obtain the maximum benefit from the exchange of information and knowledge.

## LIAISON WITH THE EMERGENCY SERVICES
The interfaces with the emergency services must be an integral part of the emergency response and as such are normally clearly defined in the emergency procedures.

## INTERFACES WITH THE MEDIA AND THE PUBLIC
One of the most precious attributes of a company is its good name and reputation. Thus it is essential that its interface with the media and public is carefully

managed to nurture its image. Again this should be formalized and included in the SMS.

## MONITORING AND INTERNAL AUDITING

It is not sufficient just to establish safety policies, procedures and a line of command and accountability. The loop must be closed by a suitable monitoring system which measures the level of performance and compares it against the company standards. Managers and supervisors need to know whether or not their programmes are being effective. Measuring the number of accidents, ill health, incidents and other evidence of deficient performance is not an effective way of monitoring safety performance, especially once it is at a reasonable level, as the numbers will be quite low and subject to random peaks and troughs. It is therefore necessary to use other yardsticks which monitor achievement of objectives and compliance with standards. There are now a number of proprietary systematic safety evaluation programmes available. Thus the SMS should define the parameters which will be monitored, the frequency of checks and the company standards which will be used for comparison purposes. One aspect of this active monitoring of safety performance is inspections of the plant and facilities. Some of these inspections will be legal requirements whilst others will be part of the company safety programme. The objective is in all cases to compare the actual situation against the defined standard. The frequency should be based on the level of potential severity or frequency, or the population at risk. High potential warrants frequent inspection, whilst low potential indicates general or infrequent inspections. To be thorough the inspections should be carried out using a check list by personnel who are competent to identify the relevant hazards and assess the conditions found (see Chapter 12).

In addition, of course, the objective is to raise progressively the company standards and it is therefore necessary to carry out periodically a detailed safety audit using a suitably qualified team who are independent from the facility being audited. The audit is carried out against a specified scope of work and is aimed at measuring the current standards and actual conditions and assessing whether these are adequate or not for present and future developments.

## SYSTEMATIC REAPPRAISAL

The results of the monitoring and auditing should be evaluated periodically together with the company principles and standards to ensure that they are still relevant. Any discrepancies should be immediately attended to along with any changes to the company structure and organization. But there should also be a gradual raising of standards so that there is continual improvement in performance. In addition to these reviews there must be a regular — say, annual — review of the whole SMS to ensure that it is up to date with all the changes that occur in companies and that it reflects current practices and goals.

The objective of the reappraisal is to maintain and develop:

- the health and safety policies;
- the safety culture;
- the performance standard;
- the SMS.

## HUMAN FACTORS

There are three major factors which influence employee behaviour. These are:

- the company safety culture and climate, which should promote involvement and commitment to the company goals and standards;
- the job itself, which must not exceed the employee's capabilities nor be demeaning;
- the strengths and weaknesses of the employee, both physical and mental, which include strength, physical limitations, habits, attitudes, personality and current stresses.

It is essential to ensure that all three factors are satisfactorily matched, and the system used to achieve this is an integral part of the SMS. To reach the levels of performance aimed at it is essential that the highest levels of management are fully committed to the policies and objectives reflected in the SMS. But even then the SMS can only be effectively implemented if it is 'owned' by the personnel who will carry out the implementation. They must share in the responsibilities, be knowledgeable of their limits of authority and the criteria by which they will be judged. To achieve 'ownership' the personnel who will have to apply the SMS should help to produce it and spend time considering and effort fostering the human factors involved in the implementation, including motivation, encouragement, involvement and awareness.

## QUALITY MANAGEMENT

One of the many attributes of a well-managed enterprise is that, in addition to producing a quality product or service, it will also invariably provide a safe working environment. In a successful company safe operation is not a separate discrete function outside the mainstream activities, but an integral part of everyone's job and hence part of the management process. Well-engineered and constructed facilities, operated and supervised by a well-trained, committed team with clearly understood objectives, will result in commercial success and a safe

157

operation. As is apparent from all this, an SMS is one component of a quality management system which should be based on a national or international quality assurance standard. These systems require a systematic definition of all the actions necessary to ensure that all activities involved in a company's operation are carried out correctly. This involves planning, organization, execution, maintenance and follow-up, and ensuring that these are carried out in accordance with company policies and procedures.

But there are other reasons for companies to use an SMS. An effective system:

- reduces accidents and ill health with the consequential financial implications;
- minimizes losses from damage to property, interruption to production and downgrading of performance;
- prevents diversion of management and supervisory effort into the investigation of the incidents;
- reduces the need for expensive modifications of plant, equipment and operating methods;
- allows a systematic approach to risk reduction and control, allowing a cost-effective application of company resources;
- develops a progressive and enlightened culture which will not only promote the safety efforts but also support the efficient operation of all company activities;
- supports a quality approach to the management of the company.

An effective SMS influences every aspect of company operation as well as the management of safety:

- corporate strategy, attitudes and ethics;
- company image;
- investment decisions;
- cost and budgetary control;
- product liability and standards;
- personnel policies;
- personnel development and career planning;
- employee participation and consultation;
- employee motivation and commitment;
- operating policy and procedures;
- environmental management;
- use of information technology.

**SELF-REGULATION**

A prerequisite for self-regulation is that an operator should work by a system of internal control which depends upon the complexity, criticality and maturity of its operation. Self-regulation requires an operating company to have in place a system of policies and standards which has to be demonstrably adhered to. The issue of verification is therefore critical. Prior to the audit process and, subsequently to make it effective, goal-setting objectives which communicate the criteria for performance and encourage the concepts of measurement and improvement should be established. The process of audit can then be used to demonstrate that the SMS is being implemented. This process should convince both management and staff of the appropriateness and success of its activities. The operator will then be able to demonstrate to the regulatory authority that actual safe practices and the understanding of the philosophy of goal-setting objectives provide sufficient confidence in the industry that the level of prescriptive legislation can be reduced, ultimately leading to self-regulation.

## CONCLUSION

Remember that the main benefit in producing an SMS is the process of its production and not the document itself. The SMS is the way that the company manages safety and how well this is implemented is how the company is judged by its employees, customers, competitors and the authorities. It is therefore in a company's self interest to ensure that not only is the SMS a realistic living up-to-date document that defines the company aims and objectives, but that it is also a system that everyone understands, believes in and follows.

# EPILOGUE

If you need convincing about the value of safety, consider the findings of one large company. This company, which is amongst the leaders in safety in its industry, sustained 39 lost work day injuries in 1985. If it had only achieved the industry average performance it would have had 372 such injuries. Since the company estimates that each lost work day injury costs approximately £150,000, it saved £6 million by being better at safety than the rest of the industry. To make £6 million in profit would have required an income of £140 million in extra sales. In addition, and even more importantly, 333 employees went home uninjured.

There is no doubt that the safer the operation the greater the operating time and the more efficient the operation and thus the higher the profitability. Safety is good business.

So how is a high standard of performance in safety achieved? The UK Health and Safety Executive Accident Prevention Advisory Unit has analysed successful managements and found that they all:

- set understandable and practical goals;
- motivate and obtain commitment from the work force;
- provide realistic resources;
- instill the need for personal responsibility in management and supervision;
- evaluate achievement.

I hope that this book will help companies to achieve these goals.

# APPENDIX — CONSULTATION AND INVOLVEMENT POLICY

The management of the company believes that it is absolutely essential that all personnel contribute their knowledge and experience to the quest of achieving safe operations. The company is therefore committed to developing procedures which encourage direct and open communications of comments, concerns and opinions about safety and environmental protection by the whole work force. It has implemented a programme of meetings and encourages all personnel, company and contractor, to take part in these. This applies equally to contract staff working on operations associated with licences operated by the company.

The company recognizes that some individuals may feel shy to voicing their opinion but it asks them to overcome those feelings and put forward their ideas since they may well prevent an accident or incident or significantly further the cause of safety and environmental protection. Equally some personnel feel concerned that if they point out a potential hazard they will be viewed as holding up progress, and may even be victimized in some way. The company is totally opposed to all forms of victimization and it will not tolerate any such activities by company personnel or contractors working for it. Anyone who feels he or she is being victimized or put under pressure to ignore safety or environmental hazards should discuss the subject with the senior company person present on the facility. If acceptable solutions cannot be achieved or an unacceptable delay is apparent, the subject should be discussed withe the relevant company Department Manager or the Safety Manager.

If agreement still cannot be achieved the matter should be raised with the Managing Director.

All claims of intimidation or victimization must be communicated up the chain of command to the Managing Director after investigation of the facts by the relevant company Manager, even if the matter has been resolved. This includes contract facilities as well as company facilities.

SIGNED BY THE CHIEF EXECUTIVE

# INDEX